COMPLETE STUDY GUIDE

TO PASS HIGH ON YOUR CIVIL SERVICE TEST

FIREFIGHTER, F.D.

By Robert E. McGannon

Deputy Chief of N.Y.C. Fire Department (ret'd)

ARCO PUBLISHING, INC.

219 Park Avenue South, New York, N.Y. 10003

Seventh Edition, Fifth Printing, 1984

Published by Arco Publishing, Inc.
219 Park Avenue South, New York, N.Y. 10003

Library of Congress Cataloging in Publication Data

McGannon, Robert E
Firefighter, F.D.

Edition published in 1971 under title: Fireman, F.D.
1. Fire extinction--Examinations, questions, etc.
2. Civil service--United States--Examinations.
I. Title.
TH9157.M28 1971 628.9'25'076 80-25047
ISBN 0-668-05170-1

Printed in the United States of America

HOW TO USE THIS INDEX
Slightly bend the right-hand edge of the book. This will expose the corresponding Parts which match the index below.

PART

1

2

3

4

CONTENTS

PART THREE

PREVIOUS TEST QUESTIONS ARRANGED TO HELP YOU STUDY EACH SUBJECT THAT MAY BE EXPECTED ON THE EXAM

PART FOUR

PRACTICE AND PREPARATION FOR THE GENERAL TEST SUBJECTS OF THE EXAMINATION

A Message from Chief McGannon

In compiling this book, it was my intention to bring to people who are interested in a Fire Department Career a study guide that is frankly based upon analysis: analysis of the previous examinations; analysis of the relevant literature on fire fighting; analysis of the policies of the Civil Service Commissions. In all this I have been guided by the vast experience gained through 23 years of service and through my experience in teaching thousands of candidates for Fire Department positions.

No effort has been spared in making this Study Course a scientific instrument for study: a valid means of forecasting the probable nature of your examination. It is my sincere hope that diligent effort will be repaid by a rewarding, lifetime job.

Requirements for Firefighters differ in most communities but usually applicants for the job must be between 21 and 29 years of age. Frequently high school graduation is required but that is not always the case. The physical test is always a rigid one, as might be expected from the nature of the job. Candidates should be in good physical condition. Specific requirements for Firefighters can always be obtained from the application form of the Civil Service Commission in the city that seeks candidates.

Robert E. McGannon
(Deputy Chief of Department, Ret.)

WHAT THIS BOOK WILL DO FOR YOU

ARCO Publishing, Inc. has followed testing trends and methods ever since the firm was founded in 1937. We specialize in books that prepare people for tests. Based on this experience, we have prepared the best possible book to help *you* score high.

To write this book we carefully analyzed every detail surrounding the forthcoming examination . . .

- the job itself
- official and unofficial announcements concerning the examination
- all the previous examinations, many not available to the public
- related examinations
- technical literature that explains and forecasts the examination

CAN YOU PREPARE YOURSELF FOR YOUR TEST?

You want to pass this test. That's why you bought this book. Used correctly, your "self-tutor" will show you what to expect and will give you a speedy brush-up on the subjects tested in your exam. Some of these are subjects not taught in schools at all. Even if your study time is very limited, you should:

- Become familiar with the type of examination you will have.
- Improve your general examination-taking skill.
- Improve your skill in analyzing and answering questions involving reasoning, judgment, comparison, and evaluation.
- Improve your speed and skill in reading and understanding what you read—an important part of your ability to learn and an important part of most tests.

This book will tell you exactly what to study by presenting in full every type of question you will get on the actual test.

This book will help you find your weaknesses. Once you know what subjects you're weak in you can get right to work and concentrate on those areas. This kind of selective study yields maximum test results.

This book will give you the *feel* of the exam. Almost all our sample and practice questions are taken from actual previous exams. On the day of the exam you'll see how closely this book follows the format of the real test.

This book will give you confidence *now,* while you are preparing for the test. It will build your self-confidence as you proceed and will prevent the kind of test anxiety that causes low test scores.

This book stresses the multiple-choice type of question because that's the kind you'll have on your test. You must not be satisfied with merely knowing the correct answer for each question. You must find out why the other choices are incorrect. This will help you remember a lot you thought you had forgotten.

After testing yourself, you may find that you are weak in a particular area. You should concentrate on improving your skills by using the specific practice sections in this book that apply to you.

THE KIND OF WORK YOU MAY BE DOING

FIREFIGHTERS

Nature of the Work

Firefighters help protect us from a hazard that costs thousands of lives and millions of dollars in property damage each year. Without their services, the loss of life and property from fires would be even greater. This statement gives information about firefighters who are full-time, paid employees of city and town fire departments. It does not cover part-time volunteer firefighters and others who serve only when the alarm signals that they are needed at a fire.

During their hours on duty at the fire station, firefighters must be prepared at a moment's notice, to rush to a fire and handle any emergency that occurs. Because firefighting is dangerous and complicated, it requires teamwork and must be well organized. At every fire, firefighters perform specific jobs assigned to them by a commanding officer; they may connect hose lines to hydrants, operate a pressure pump, position ladders, or perform some other duty. Furthermore, depending on the judgment of the officer in charge, the assigned duties of individual firefighters may be changed several times while the company is in action. Under emergency conditions firefighters are often called on to use their own initiative and judgment. Firefighters, therefore must be proficient in many different kinds of firefighting activities, as well as being capable of helping people to safety, administering first aid, and taking care of other emergencies.

Fire prevention is another important responsibility of city fire departments. Specially trained personnel inspect factories, theaters, and other public buildings for conditions that might cause a fire and for compliance with local regulations relating to fire escapes, fire doors, storage of flammable materials, and other possible hazards. Educating the public about fire prevention and safety measures is also a part of the firefighter's job. Frequently, they speak on this subject before school assemblies and civic groups. In many communities, they regularly inspect private homes, at the owner's request, in an effort to prevent fires by pointing out possible hazards to homeowners.

Between alarms, firefighters spend considerable time at their local stations, improving their knowledge of firefighting and doing maintenance work. They also participate in practice drills, clean and lubricate firefighting equipment, stretch hoses to dry, stand watch at fire alarm instruments, and verify and record alarms.

Places of Employment

The number of full-time firefighters employed by city fire departments is estimated at more than 180,000. In addition, thousands of paid "call men and women" and hundreds of thousands of part-time volunteer firefighters are organized in small towns and rural communities throughout the nation to help fight fires. A few very large cities have several thousand firefighters; some small cities have fewer than 25.

Training, Other Qualifications, and Advancement

To become eligible for an appointment as a firefighter, an applicant must pass a written intelligence test, a medical examination, and tests of strength, physical stamina, and agility, as specified by local civil service regulations. In most communities, these examinations are open only to men and women who are at least 18 years of age, meet certain height and weight requirements, and have a high school education. The men and women who receive the highest grades on their examinations have the best chances for appointment. Extra credit usually is given for military service. Experience gained as a volunteer firefighter or through firefighting training in the Armed Forces also may improve an applicant's chances for appointment.

As a rule, beginners in large fire departments are given training for several weeks at the city's fire school. Through classroom instruction and practice drills, the recruits study such fundamentals as firefighting techniques, local building codes, fire prevention, and first aid; and learn about the use of axes, chemical extinguishers, ladders, and other firefighting equipment. Upon completion of this training, they are assigned to local fire companies. Opportunities for promotion are good in most fire departments. As firefighters gain experience, they may be advanced to progressively higher ratings, and, after 5 to 10 years or more of service, become eligible for promotion to the grade of lieutenant. The line of further promotion is usually to captain, then battalion chief, assistant chief, and finally to chief. Chances for advancement generally depend upon each candidate's position on the promotion list, as determined by the written examination, performance as a firefighter, and seniority. Throughout their service, many firefighters continue to study fire prevention and related subjects to improve their performance on the job and prepare for promotional examinations. Programs conducted by many state governments and city fire departments throughout the country provide training of this kind for tens of thousands of firefighters each year. Some universities offer courses in fire engineering.

Among the important personal qualities of firefighters are mental alertness, courage, mechanical aptitude, and endurance. Initiative and good judgment are extremely important, because firefighters often must make quick decisions as situations change while companies are in action. Leadership qualities are valuable assets for officers, who have the responsibility for establishing and maintaining a high degree of discipline and efficiency, as well as planning and directing the activities of the firefighters in their companies.

Employment Outlook

Several thousand openings for firefighters occur each year. Many openings will arise from the need to replace those who retire, or otherwise leave. The replacement rate for firefighters is higher than for most occupations, largely because retirement is often permitted at an earlier age than in many other occupations. New jobs also will become available as city fire departments enlarge their staffs and as paid departments replace volunteer fire companies in smaller, growing communities. In addition, some openings probably will be created as city fire departments continue to shorten the scheduled hours that individual firefighters are on duty.

The number of young people who qualify for firefighter jobs in large cities usually is greater than the number of job openings, even though the written examination and physical requirements eliminate many applicants. Competition among candidates is apt to be keen since employment in this occupation is very stable.

The number of firefighters is expected to increase rapidly to meet the needs for fire protection in growing urban communities. As cities become more crowded, however, officials will give more emphasis to activities associated with fire prevention, and many firefighters will spend a greater amount of their time inspecting buildings for compliance with fire regulations and participating in fire prevention campaigns.

Earnings and Working Conditions

Firefighters in the larger cities usually receive the highest starting salaries. Fire chiefs receive average salaries ranging from $15,000 a year in the smaller cities to $40,000 a year in cities that have populations over 250,000.

Practically all fire departments furnish pay allowances for protective firefighting clothing (helmets, boots, and rubber coats) and many also provide dress uniforms.

Firefighters' scheduled hours range from 40 hours a week in some cities to 60 hours in others; the national average workweek is about 56 hours. The scheduled workweek in metropolitan centers having large fire departments tends to be considerably shorter than in small communities, but recent developments in collective bargaining have resulted in a general trend toward reducing the workweek of the firefighter.

In addition to their scheduled hours, firefighters must work as many extra hours as necessary to bring a fire under control. When overtime is worked, most city fire departments either give compensatory time off or extra pay for the additional hours.

The job of a firefighter involves risk of life or injury from sudden cave-ins of floors or toppling walls, as well as hazards associated with exposure to flames, smoke, and bad weather. In fighting fires in industrial establishments, firefighters may come in contact with poisonous, flammable, and explosive gases and chemicals.

Firefighters generally are covered by liberal pension plans, many of which provide for retirement at half pay at age 50 after 25 years of service, or at any age if disabled in the line of duty. Firefighters also receive regular paid vacations. Provisions for sick leave usually are very liberal; health and surgical benefit plans are offered in many fire departments; and compensation also is provided for firefighters injured in the line of duty. Most fire departments either allow paid holidays—ranging up to 11 or more a year—or time off for working on holidays.

Firefighters are granted unlimited sick leave because of the dangerous and physically taxing duties they are required to perform. Full pay is authorized during the entire period they have been evaluated as being medically unfit to perform their duties by a Medical Officer. However, a disability that is evaluated as being of a permanent nature may result in a premature retirement. If the disability is construed to be service connected, ¾ pay is usually granted. If the disability is evaluated as not having resulted from the firefighter's duties, limited retirement benefits are authorized.

PART 1

Applying and Studying For Your Job

FIREFIGHTER, F. D.

TYPICAL JOB ANNOUNCEMENT REQUIREMENTS

Before you can be appointed as a firefighter, you must meet the following requirements and you must receive at least 70% on the written multiple-choice test and at least 70% on the physical test.

AGE REQUIREMENTS: To qualify for *the examination* a person shall have passed his eighteenth birthday, but not his twenty-ninth birthday on the date of the written test. However, no person shall be *appointed* unless he shall have attained his twenty-first birthday.

EXCEPTIONS: All persons who were engaged in military duty, as defined in Section 243 of the Military Law, subsequent to July 1, 1940, may deduct the length of time, not exceeding a total of six years which they spent in military service, from their actual age in determining their eligibility. At the time of investigation, applicants will be required to present proof of date of birth by transcript of record of the Bureau of Vital Statistics or other satisfactory evidence.

MINIMUM REQUIREMENTS: Persons lacking a high school diploma are permitted to apply for and take this examination. However, at the time of appointment eligibles must possess a high school diploma or present evidence of having passed an examination for the high school equivalency diploma or United States Armed Forces G.E.D. certificate with a score of at least 35 on each of the five tests and an overall score of at least 225 in the examination for the diploma or certificate.

Proof of the possession of the required diploma or certificate must be presented at the time of appointment interview and to the Department of Personnel at the time of investigation.

Eligibles unable to meet this requirement when called for their appointment interview will remain on the eligible list in accordance with their rank position but will not be certified for appointment until they have met this requirements.

Proof of good character will be a prerequisite to appointment. Persons convicted of a felony are not eligible for positions in the uniformed forces of the Fire Department. In addition, the rules of the Department of Personnel provide that no person who has been dishonorably discharged from the Armed Forces shall be examined or certified or appointed as a Firefighter.

Applicants must be not less than 5 feet 4 inches (bare feet) in height and must approximate normal weight for height. Height will be tested after strenuous physical activity.

Required vision: 20/40 for each eye separately without glasses; 20/30 both eyes together without glasses.

LICENSE REQUIREMENT: At the time of appointment, applicants must have a valid New York State motor vehicle driver's license.

TESTS: The physical test will be designed to test the candidates' capacity to perform the physical aspects of the work of a Firefighter. It will consist of seven sub-tests, as follows: mile run-walk; dummy carry; agility; hand grip; standing broadjump; ledge balance and flexed arm hang.

Eligibles will be required to pass a qualifying medical test prior to appointment. Candidates will be required to meet the medical and physical standards of their city Department of Personnel.

MEDICAL STANDARDS: Detailed medical standards and detailed physical test requirements will be established and posted before the date of the written test.

Medical evidence to allow participation in the physical test may be required and the Department of Personnel reserves the right to exclude from the physical test any candidates who, upon examination of such evidence, are apparently medically unfit.

Candidates will take the physical test at their own risk of injury, although every effort will be made to safeguard them.

PHYSICAL TEST: WEIGHT 50, 70% REQUIRED.

SUBTEST I: MILE RUN-WALK—ONE trial only.

The candidates will be required to run or use a combination of running and walking to complete a one mile distance around a track, as quickly as they can. The scoring is as follows:

Minutes and Seconds to Complete	4:50 or less	4:50-4:56	4:57-5:03	5:04-5:11	5:12-5:19	5:20-5:27	5:28-5:36	5:37-5:45	5:46 5:54
Weighted Percent	200	195	190	185	180	175	170	165	160
Minutes and Seconds to Complete	5:55-6:03	6:04-6:12	6:13-6:21	6:22-6:30	6:31-6:39	6:40-6:48	6:49-6:56	6:57-7:04	7:05-7:11
Weighted Percent	155	150	145	140	135	130	125	120	115
Minutes and Seconds to Complete	7:12-7:18	7:19-7:24	7:25-7:30	7:30 or More					
Weighted Percent	110	105	100	Zero					

SUBTEST II: DUMMY CARRY—ONE trial (unless a zero penalty score, then a second and final trial allowed).

Standing with both feet behind the start line, on the signal "Go," step forward, pick up 120-pound dummy, put it on one shoulder, carry in that position up one flight of stairs, walk around marked lines, carry downstairs, bend on one knee and return dummy to mat under control. Dummy must be kept off ground while carrying. Score is time to complete. (Note: A 2-second penalty is added to total time score if dummy is dropped out of control after carrying up and down stairs. Dropping dummy anywhere else during test results in zero score.) The scoring is as follows:

Score in Seconds	18/less	19	20	21	22	23	24	25	26	27	28	29	30
Weighted Percent	150	145	141	137	133	129	125	121	117	113	109	105	100
Score in Seconds	31	32	33	34	35								
Weighted Percent	95	90	85	80	75								

Still unfinished after 35 seconds:

Lower dummy to ground	-70
On upper stair landing or returning downstairs	-35
Cannot lift or going up stairs	-Zero

SUBTEST III: AGILITY—THE BETTER OF *TWO* TRIALS

From the supine position, feet together, hands at sides, on signal to "Go," rise and run 9 feet to the 5-foot wall and scale it; run 11½ feet to a maze of obstacles and dodge through; run about 15 feet and crawl through the first "window"; mount the 30-foot horizontal ladder and crawl across it keeping one hand on each outside rail; crawl through the second "window"; run about 20 feet to the 8-foot wall and scale it; run around one obstacle; sprint 45 feet to finish line. Score is time to complete the course. (Note: a penalty of 2 seconds added to total time will be assessed each time a foot touches the floor on the outside of the ladder while crawling across it. A 2-second penalty will be assessed and added to total time if both feet fail to land beyond the last rung of the ladder and on the side of the ladder farthest away from the second "window." Use of the iron supporting rods to climb the 8-foot wall or running out of the course without retracking and continuing properly within the time limit will result in credit only for previous properly completed obstacles.) The scoring is as follows:

Score in Seconds	22 or less	23	24	25	26	27	28	29	30	31	32	33	34
Weighted Percent	250	245	240	235	229	223	217	211	205	199	193	187	181
Score in Seconds	35	36	37	38	39	40	41	42	43				
Weighted Percent	175	169	163	157	151	145	139	132	125				

Still unfinished after 43 seconds:

Sprint	-118
8-foot wall	-59
"Window," ladder, maze, or 5-foot wall	-Zero

SUBTEST IV: HAND GRIP—THE BETTER OF *TWO* TRIALS

While standing, hold dynamometer in preferred hand, with palm around the top bar and the fingers curled around the bottom bar, arm extended downward away from the side of the body and with the palm facing the side of the body, squeeze the dynamometer as sharply and as steadily as possible. Score is highest possible dynamometer reading obtained by one squeeze. The scoring is as follows:

Score in Kilograms	70 or more	69-65	64-60	59-55	54-50	49-46	45-42	41-40
Weighted Percent	100	95	90	85	80	75	70	65
Score in Kilograms	39-38	37-36	35-34	Less than 34				
Weighted Percent	60	55	50	Zero				

(*NOTE:* 1 kilogram = 2.2 pounds. There will be a rest period of one minute between trials.)

SUBTEST V: STANDING BROAD JUMP—THE BETTER OF *TWO* TRIALS

With both feet behind the start line, jump forward as far as possible without falling backwards. Arms may be used free-style during jump. Score is the greater distance jumped as measured from the point where the rear-most heel lands from the start line. (*NOTE:* Falling backwards when landing results in a zero score for that trial.) The scoring is as follows:

Score in Feet and Inches	8'8" or Better	8'5" or Better	8'2" or Better	7'10" or Better	7'6" or Better	7'2" or Better	6'10" or Better
Weighted Percent	100	95	90	85	80	75	70

Score in Feet and Inches	6'8" or Better	6'6" or Better	6'4" or Better	6'2" or Better	Less Than 6'2"
Weighted Percent	65	60	55	50	Zero

SUBTEST VI: LEDGE BALANCE—THE BETTER OF *TWO* TRIALS

Wearing an "oxygen unit" consisting of a harness and weight on the back (weight: 26 lbs.), start with both feet facing the wall, within the one foot area marked off on the 30-foot "ledge" that is 2½ inches wide, and 6 inches from the wall. On the signal "Go," side step to the right as quickly as possible to the opposite side of the "ledge" without falling off; touch the side board with the right foot and make sure that both feet are within the one-foot marked off area; then side step to the left as quickly as possible, without falling, until both feet are again within the one-foot marked off starting area. Score is the time to move across "ledge" and back again without falling. (*NOTE:* Falling off "ledge" constitutes a zero score for that trial.) The scoring is as follows:

Score in Seconds	9 or less	10	11	12	13	14	15	16	17	18	19	20	21
Weighted Percent	100	97	94	90	86	82	78	74	70	65	60	55	50

Score in Seconds	More than 21
Weighted Percent	Zero

SUBTEST VII: FLEXED ARM HANG—ONE trial only

Mount the ladder to a height such that the chin is even with the horizontal bar; use a firm, overhand grip on the bar and when ready, nod, and the ladder will be taken away, and timing will begin. Hang onto bar as long as possible. The test will be over when arms are completely extended. Score is time before arms are completely extended. The scoring is as follows:

Score in Minutes and Seconds	2:20 or More	2:19-2:10	2:09-2:00	1:59-1:50	1:49-1:40	1:39-1:30	1:29-1:20	1:19-1:15	1:14-1:10
Weighted Percent	100	95	90	85	80	75	70	65	60

Score in Minutes and Seconds	1:09-1:05	1:04-1:00	Less than 1:00
Weighted Percent	55	50	Zero

THE APPLICATION FORM

FILLING OUT THE APPLICATION FORMS

If you are satisfied that you meet the requirements listed in the announcement, the next step is to fill out the application forms. At first you may have to fill out only a small card, but sooner or later you will have to fill out a 2- or a 4-page personal qualifications statement. It is very important to do this carefully.

Answer every question in the statement. If you don't, the interagency board of examiners will have to write to you to get the missing information. This will take time and delay action on your qualifications statement. You will ordinarily be given only one opportunity to send in this information. If you don't reply promptly, your statement will be canceled.

For many positions, written tests are not required, and civil service examiners rate applicants on their training and experience. Tell the whole story. You can't get credit for experience and training which you don't claim on your statement.

Follow the instructions in the announcement as to when and where to send your statement. Be sure to send it to the right office before the closing date.

WHAT TO EXPECT ON THE APPLICATION FORM

While there are minor differences in the questions asked, and the order in which they're presented, by and large this is what you can expect:

Name of examination, or kind of position applied for. This information appears in large type at the top of job announcements. All you have to do is to copy it fully.

Optional subject (if mentioned in announcement)—Many examination announcements have options, that is, a choice of the kind of work you want to do. To answer this question you write the specific option of the job you are applying for, again copying it from the announcement form.

AGE AND PLACE OF BIRTH

Your place of birth.

Your date of birth (month, day, year).

The Government makes no distinction between people born abroad and those born in the United States in selecting workers for government positions. However, job holders in the competitive service must be citizens.

THE "OMNIBUS" QUERY

We now reach the "omnibus" query. This consists of six parts and deals with specific information concerning certain preferences of yours about salary, length and location of employment. The information you provide in response to this question will enable the government to offer you a wider or narrower range of jobs and provide civil

service authorities with an idea of your feelings about a short-term job if one should come up.

EXPERIENCE

This section consists of a series of blocks which you must fill in with the details of the positions you have held. The details which the Commission wants include: dates of employment, place of employment, name and address of employer, kind of business or organization (for example, dress manufacturing, wholesale grocer, service station, insurance agency), the number and kind of employees you supervised if you were in a supervisory position, the name and title of your immediate supervisor, the exact title of your position, the salary you earned when you started and your final salary, a description of your work. The government also wants to know your reason for desiring to change employment and the reasons for leaving your former positions.

HAVE YOU ANY SPECIAL APTITUDES?

Special Qualifications and Skills: This section is included so that you may cite any special qualifications not covered elsewhere in the application blank. It asks for a list of special skills you possess, machines and equipment you can use, such as operation of shortwave radio, multilith, comptometer, keypunch, turret lathe, scientific or professional devices. Here, too, is a space in which you can list your typing and shorthand speed.

TECHNIQUES OF STUDY AND TEST-TAKING

Although a thorough knowledge of the subject matter is the most important factor in succeeding on your exam, the following suggestions could raise your score substantially. These few pointers will give you the strategy employed on tests by those who are most successful in this not-so-mysterious art. It's really quite simple. Do things right . . . right from the beginning. Make these successful methods a habit. Then you'll get the greatest dividends from the time you invest in this book.

PREPARING FOR THE EXAM

1. *Budget your time.* Set aside definite hours each day for concentrated study. Keep to your schedule.

2. *Study alone.* You will concentrate better when you work by yourself. Keep a list of questions you cannot answer and points you are unsure of to talk over with a friend who is preparing for the same exam. Plan to exchange ideas at a joint review session just before the test.

3. *Eliminate distractions.* Disturbances caused by family and neighbor activities (telephone calls, chit-chat, TV programs, etc.) work to your disadvantage. Study in a quiet, private room.

4. *Use the library.* Most colleges and universities have excellent library facilities. Some institutions have special libraries for the various subject areas: physics library, education library, psychology library, etc. Take full advantage of such valuable facilities. The library is free from those distractions that may inhibit your home study. Moreover, research in your subject area is more convenient in a library since it can provide more study material than you have at home.

5. *Answer all the questions in this book.* Don't be satisfied merely with the correct answer to each question. Do additional research on the other choices which are given. You will broaden your background and be more adequately prepared for the "real" exam. It's quite possible that a question on the exam which you are going to take may require you to be familiar with the other choices.

6. *Get the "feel" of the exam.* The sample questions which this book contains will give you that "feel" since they are virtually the same as those you will find on the test.

7. *Take the Sample Tests as "real" tests.* With this attitude, you will derive greater benefit. Put yourself under strict examination conditions. Tolerate no interruptions while you are taking the sample tests. Work steadily. Do not spend too much time on any one question. If a question seems too difficult go to the next one. If time permits, go back to the omitted question.

8. *Tailor your study to the subject matter. Skim or scan.* Don't study everything in the same manner. Obviously, certain areas are more important than others.

9. *Organize yourself.* Make sure that your notes are in good order—valuable time is unnecessarily consumed when you can't find quickly what you are looking for.

10. *Keep physically fit.* You cannot retain information well when you are uncomfortable, headachy, or tense. Physical health promotes mental efficiency.

HOW TO TAKE AN EXAM

1. *Get to the Examination Room about Ten Minutes Ahead of Time.* You'll get a better start when you are accustomed to the room. If the room is too cold, or too warm, or not well ventilated, call these conditions to the attention of the person in charge.

2. *Make Sure that You Read the Instructions Carefully.* In many cases, test-takers lose credits because they misread some important point in the given directions—example: the *incorrect* choice instead of the *correct* choice.

3. *Be Confident.* Statistics conclusively show that high scores are more likely when you are prepared. It is important to know that you are not expected to answer every question correctly. The questions usually have a range of difficulty and differentiate between several levels of skill.

4. *Skip Hard Questions and Go Back Later.* It is a good idea to make a mark on the question sheet next to all questions you cannot answer easily, and to go back to those questions later. First answer the questions you are sure about. Do not

panic if you cannot answer a question. Go on and answer the questions you know. Usually the easier questions are presented at the beginning of the exam and the questions become gradually more difficult.

If you do skip ahead on the exam, be sure to skip ahead also on your answer sheet. A good technique is periodically to check the number of the question on the answer sheet with the number of the question on the test. You should do this every time you decide to skip a question. If you fail to skip the corresponding answer blank for that question, all of your following answers will be wrong.

Each student is stronger in some areas than in others. No one is expected to know all the answers. Do not waste time agonizing over a difficult question because it may keep you from getting to other questions that you can answer correctly.

5. *Guess If You Are Not Sure.* No penalty is given for guessing when these exams are scored. Therefore, it is better to guess than to omit an answer.

6. *Mark the Answer Sheet Clearly.* When you take the examination, you will mark your answers to the multiple-choice questions on a separate answer sheet that will be given to you at the test center. If you have not worked with an answer sheet before, it is in your best interest to become familiar with the procedures involved. Remember, knowing the correct answer is not enough! If you do not mark the sheet correctly, so that it can be machine-scored, you will not get credit for your answers!

In addition to marking answers on the separate answer sheet, you will be asked to give your name and other information, including your social security number. As a precaution bring along your social security number for identification purposes.

Read the directions carefully and follow them exactly. If they ask you to print your name in the boxes provided, write only one letter in each box. If your name is longer than the number of boxes provided, omit the letters that do not fit. Remember, you are writing for a machine; it does not have judgment. It can only record the pencil marks you make on the answer sheet.

Use the answer sheet to record all your answers to questions. Each question, or item, has four or five answer choices labeled (A), (B), (C), (D), (E). You will be asked to choose the letter that stands for the best answer. Then you will be asked to mark your answer by blackening the appropriate space on your answer sheet. Be sure that each space you choose and blacken with your pencil is *completely* blackened. The machine will "read" your answers in terms of spaces blackened. Make sure that only one answer is clearly blackened. If you erase an answer, erase it completely and mark your new answer clearly. The machine will give credit only for clearly marked answers. It does not pause to decide whether you really meant (B) or (C).

Make sure that the number of the question you are being asked on the question sheet corresponds to the number of the question you are answering on the answer sheet. It is a good idea to check the numbers of questions and answers frequently. If you decide to skip a question, but fail to skip the corresponding answer blank for that question, all your answers after that will be wrong.

7. *Read Each Question Carefully*. The exam questions are not designed to trick you through misleading or ambiguous alternative choices. On the other hand, they are not all direct questions of factual information. Some are designed to elicit responses that reveal your ability to reason, or to interpret a fact or idea. It's up to you to read each question carefully, so you know what is being asked. The exam authors have tried to make the questions clear. Do not go astray looking for hidden meanings.

8. *Don't Answer Too Fast*. The multiple-choice questions which you will meet are not superficial exercises. They are designed to test not only your memory but also your understanding and insight. Do not place too much emphasis on speed. The time element is a factor, but it is not all-important. Accuracy should not be sacrificed for speed.

9. *Materials and Conduct at the Test Center*. You need to bring with you to the test center your Admission Form, your social security number, and several No. 2 pencils. Arrive on time as you may not be admitted after testing has begun. Instructions for taking the tests will be read to you by the test supervisor and time will be called when the test is over. If you have questions, you may ask them of the supervisor. Do not give or receive assistance while taking the exams. If you do, you will be asked to turn in all test materials and told to leave the room. You will not be permitted to return and your tests will not be scored.

HOW TO BE A MASTER TEST TAKER

SCORING PAPERS BY MACHINE

A typical machine-scored answer sheet is shown below, reduced from the actual size of 8¼ x 11 inches. Since it's the only one that reaches the office where papers are scored, it's important that the blanks at the top be filled in completely and correctly.

You'll have to mark your answers on one of these sheets. Consequently, we've made it possible for you to practice with them throughout this book.

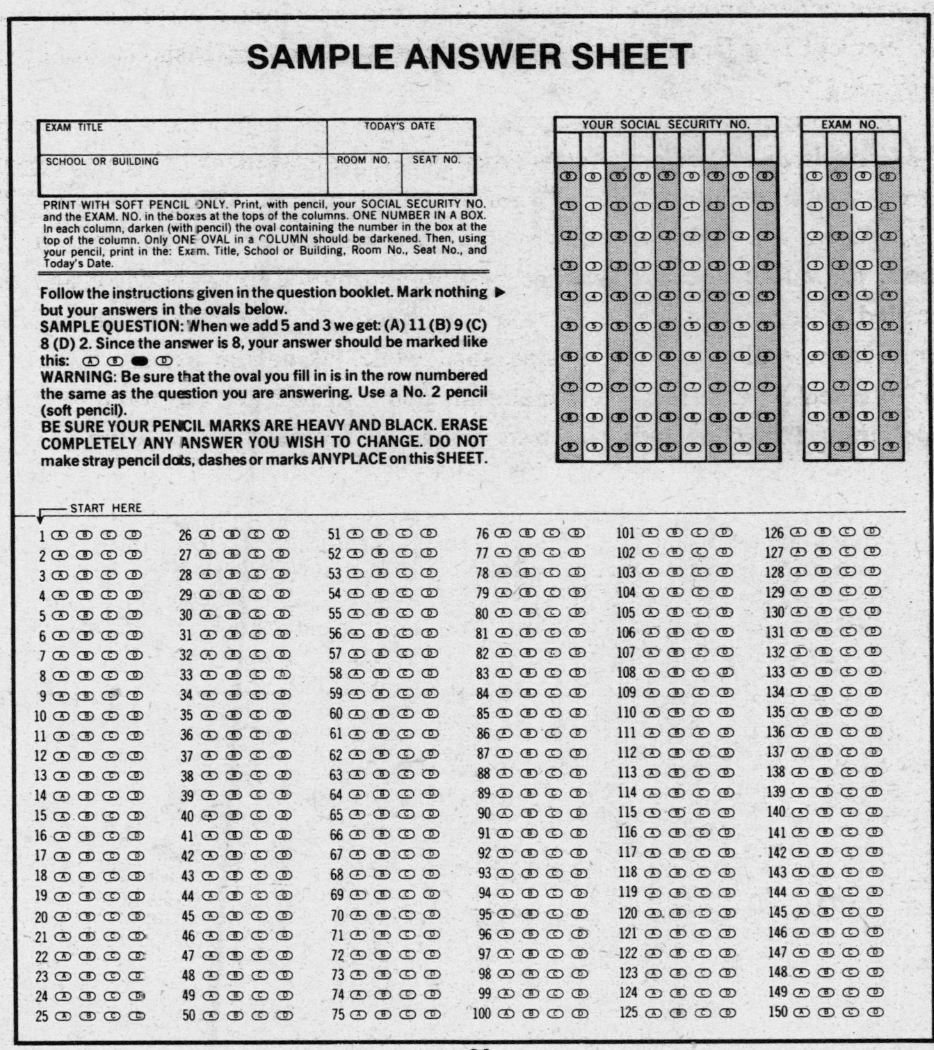

FOLLOW DIRECTIONS CAREFULLY

It's an obvious rule, but more people fail for breaching it than for any other cause. By actual count there are over a hundred types of directions given on tests. You'll familiarize yourself with many of them in the course of this book. And you'll also learn not to let your guard down in reading them, listening to them, and following them. Right now, before you plunge in, we want to be sure that you have nothing to fear from the answer sheet and the way in which you must mark it; from the most important question forms and the ways in which they are to be answered.

HERE'S HOW TO MARK YOUR ANSWERS ON MACHINE-SCORED ANSWER SHEETS:

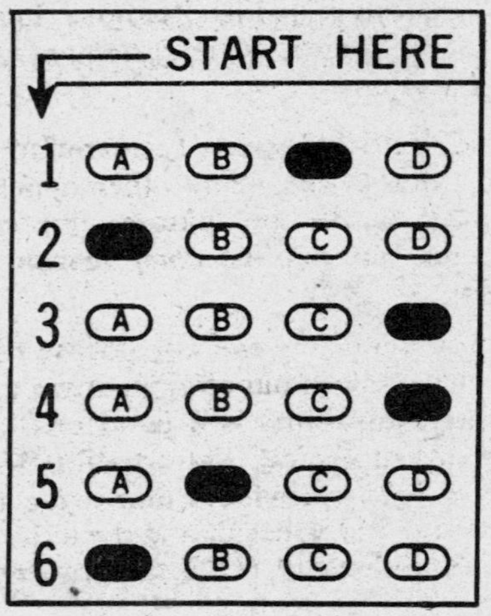

Make only ONE mark for each answer. Additional and stray marks may be counted as mistakes. In making corrections, erase errors COMPLETELY. Make glossy black marks.

(a) Each pencil mark must be heavy and black. Light marks should be retraced with the special pencil.

(b) Each mark must be in the space between the pair of dotted lines and entirely fill this space.

(c) All stray pencil marks on the paper, clearly not intended as answers, must be completely erased.

(d) Each question must have only one answer indicated If multiple answers occur, all extraneous marks should be thoroughly erased. Otherwise, the machine will give you *no* credit for your correct answer.

MULTIPLE CHOICE METHODS

Multiple choice questions offer you four or five possible answers. Your job is to select *only* the *best* answer. Even the incorrect answers are frequently *partly* correct. They are inserted to force you to think . . . and prove that you know the right answer.

THE BEST WAY TO ANSWER MULTIPLE CHOICE QUESTIONS:

1. Read each question carefully to see what the examiner is after. Re-read always.
2. Forget about answers that are clearly wrong.
3. You should be suspicious of choices which contain broad statements using words like "all," "always," "never" and "only."

4. A well-constructed multiple choice item will avoid obviously incorrect choices. The good examiner will try to write a cluster of answers, all of which are plausible. Use the clue words to help yourself pick the *most* correct answer.

5. In the case of items where you are doubtful of the answer, you might be able to bring to bear the information you have gained from previous study. This knowledge might be sufficient to indicate that some of the suggested answers are not so plausible. Eliminate such answers from further consideration.

6. Then concentrate on the remaining suggested answers. The fewer answers you have to consider, the better your chances of getting the item right.

7. If the item is in the form of an incomplete statement, it sometimes helps to try to complete the statement before you look at the suggested answers. Then see if the way you have completed the statement corresponds with any of the answers provided. If one is found, it is likely to be the correct one.

8. Use your head! Make shrewd inferences. Sometimes, with a little thought, and the information that you have, you can reason out the answer. We're suggesting a method of intelligent guessing in which you can become quite expert with a little practice. It's a useful method that may help you with some debatable answers.

debatable answers.

NOW, LET'S TRY THESE METHODS OUT ON A SAMPLE MULTIPLE-CHOICE QUESTION.

1. Leather is considered the best material for shoes chiefly because
   (A) it is waterproof
   (B) it is quite durable
   (C) it is easily procurable
   (D) it is flexible and durable
   (E) it can be easily manufactured in various styles.

Here we see that every one of the answer statements is plausible: leather is waterproof if treated properly; it is relatively durable; it is relatively easily procurable; it bends and is shaped easily, and is, again, durable; it constantly appears in various styles of shoes and boots.

However, we must examine the question with an eye toward identifying the key phrase which is: *best* for shoes *chiefly*.

Now we can see that (A) is incorrect because leather is probably not the *best* material for shoes, simply because it is waterproof. There are far better waterproof materials available, such as plastics and rubber. In fact, leather must be treated to make it waterproof. So by analyzing the key phrase of the question we can eliminate (A).

(B) seems plausible. Leather is durable and durability is a good quality in shoes. This answer may be correct.

(C) Leather is comparatively easy to procure; but would that make it *best* for shoes? And would that be the *chief* reason why it is used for making shoes? Although the statement in itself is quite true, it does not fit the key phrase of the question and we must, reluctantly, eliminate it.

(D) is a double-barreled statement. One part, the durability, has been suggested in (B) above. Leather is also flexible, so both parts of the statement would seem to fit the question.

(E) It is true that leather can be manufactured in various styles, but so can many other materials. Again, going back to the key phrase, this could be considered one, but not the *chief* reason why it is *best* for shoes.

So, by carefully analyzing the *key* phrase of the question we have narrowed our choices down to (B) and (D). Although durability is a good quality in shoes, it is only one of several. The combination of durability with flexibility provides a more exact definition of the qualities of leather that make it the best material for shoes. Therefore, (D) is the best answer to this question.

The same question, by slightly altering the answer choices, can also call for a *negative* response. Here, even more so, the identification of the key phrase becomes vital in finding the correct answer. Suppose the question and its responses were worded thus:

2. Leather is considered the best material for shoes chiefly because
   (A) it is waterproof
   (B) it is easily colored
   (C) it is easily procurable
   (D) it can easily be manufactured in various styles
   (E) none of these.

We can see that the prior partially correct answer (B) has now been changed, and the doubly-correct answer eliminated. Instead we have a new response possibility (E), "none of these."

We have analyzed three of the choices previously and have seen the reason why none of them is the *chief* reason why leather is considered the *best* material for shoes. The two new elements are (B) "easily colored," and (E) "none of these."

If you think about it, leather *can* be easily colored and often is, but this would not be the chief reason why it is considered *best.* Many other materials are just as easily dyed. So we must come to the conclusion that *none* of the choices is *completely* correct—none fits the key phrase. Therefore, the question calls for a negative response (E).

We have now seen how important it is to identify the key phrase. Equally, or perhaps even more important, is the identifying and analyzing of the key *word*—the qualifying word—in a question. This is usually, though not always, an adjective or adverb. Some of the key words to watch for are: *most, best, least, highest, lowest, always, never, sometimes, most likely, greatest, smallest, tallest, average, easiest, most nearly, maximum, minimum, chiefly, mainly, only, but* and *or.* Identifying these key words is usually half the battle in understanding and, consequently, answering all types of exam questions.

Rephrasing the Question

It is obvious, then, that by carefully analyzing a question, by identifying the key phrase and its key words, you can usually find the correct answer by logical deduction and, often, by elimination. One other way of examining, or "dissecting," a question is to restate or rephrase it with each of the suggested answer choices integrated into the question.

For example, we can take the same question and rephrase it.

(A) The chief reason why leather is considered the best material for shoes is that it is waterproof.

or

(A) Because it is waterproof, leather is considered the best material for shoes.

or

(A) Chiefly because it is waterproof, leather is considered the best material for shoes.

It will be seen from the above three new versions of the original statement and answer that the question has become less obscure because it has been, so to speak, illuminated from different angles. It becomes quite obvious also in this rephrasing that the statement (A) is incorrect, although the *original* phrasing of the question left some doubt.

The rules for understanding and analyzing the key phrase and key words in a question, and the way to identify the *one* correct answer by means of intelligent analysis of the important question-answer elements, are basic to the solution of all the problems you will face on your test.

In fact, perhaps the *main* reason for failing an examination is failure to *understand the question.* In many cases, examinees *do* know the answer to a particular problem, but they cannot answer correctly because they do not understand it.

2

PART 2

Sample Exams for Practice

DIRECTIONS FOR ANSWERING QUESTIONS

For each question read all the choices carefully. Then select that answer which you consider correct or most nearly correct. Write the letter preceding your best choice next to the question

Should you want to answer on the kind of answer sheet used on machine-scored examinations, we have provided several such facsimiles. Tear one out if you wish, and mark your answers on it . . . just as you would do on an actual exam

In machine-scored examinations you should record all your answers on the answer sheet provided. Don't make the mistake of putting answers on the test booklet itself.

On some machine-scored exams you are instructed to "place no marks whatever on the test booklet."

In other examinations you may be instructed to mark your answers in the test booklet. In such cases you should be careful that no other marks interfere with the legibility of your answers.

It is always best NOT to mark your booklet unless you are sure that it is permitted.

SAMPLE ANSWER SHEET

EXAM TITLE	TODAY'S DATE	
SCHOOL OR BUILDING	ROOM NO.	SEAT NO.

PRINT WITH SOFT PENCIL ONLY. Print, with pencil, your SOCIAL SECURITY NO. and the EXAM. NO. in the boxes at the tops of the columns. ONE NUMBER IN A BOX. In each column, darken (with pencil) the oval containing the number in the box at the top of the column. Only ONE OVAL in a COLUMN should be darkened. Then, using your pencil, print in the: Exam. Title, School or Building, Room No., Seat No., and Today's Date.

Follow the instructions given in the question booklet. Mark nothing but your answers in the ovals below.
SAMPLE QUESTION: When we add 5 and 3 we get: (A) 11 (B) 9 (C) 8 (D) 2. Since the answer is 8, your answer should be marked like this: (A) (B) ● (D)
WARNING: Be sure that the oval you fill in is in the row numbered the same as the question you are answering. Use a No. 2 pencil (soft pencil).
BE SURE YOUR PENCIL MARKS ARE HEAVY AND BLACK. ERASE COMPLETELY ANY ANSWER YOU WISH TO CHANGE. DO NOT make stray pencil dots, dashes or marks ANYPLACE on this SHEET.

YOUR SOCIAL SECURITY NO.								
0	0	0	0	0	0	0	0	0
1	1	1	1	1	1	1	1	1
2	2	2	2	2	2	2	2	2
3	3	3	3	3	3	3	3	3
4	4	4	4	4	4	4	4	4
5	5	5	5	5	5	5	5	5
6	6	6	6	6	6	6	6	6
7	7	7	7	7	7	7	7	7
8	8	8	8	8	8	8	8	8
9	9	9	9	9	9	9	9	9

EXAM NO.			
0	0	0	0
1	1	1	1
2	2	2	2
3	3	3	3
4	4	4	4
5	5	5	5
6	6	6	6
7	7	7	7
8	8	8	8
9	9	9	9

START HERE

1 (A) (B) (C) (D)	26 (A) (B) (C) (D)	51 (A) (B) (C) (D)	76 (A) (B) (C) (D)	101 (A) (B) (C) (D)	126 (A) (B) (C) (D)
2 (A) (B) (C) (D)	27 (A) (B) (C) (D)	52 (A) (B) (C) (D)	77 (A) (B) (C) (D)	102 (A) (B) (C) (D)	127 (A) (B) (C) (D)
3 (A) (B) (C) (D)	28 (A) (B) (C) (D)	53 (A) (B) (C) (D)	78 (A) (B) (C) (D)	103 (A) (B) (C) (D)	128 (A) (B) (C) (D)
4 (A) (B) (C) (D)	29 (A) (B) (C) (D)	54 (A) (B) (C) (D)	79 (A) (B) (C) (D)	104 (A) (B) (C) (D)	129 (A) (B) (C) (D)
5 (A) (B) (C) (D)	30 (A) (B) (C) (D)	55 (A) (B) (C) (D)	80 (A) (B) (C) (D)	105 (A) (B) (C) (D)	130 (A) (B) (C) (D)
6 (A) (B) (C) (D)	31 (A) (B) (C) (D)	56 (A) (B) (C) (D)	81 (A) (B) (C) (D)	106 (A) (B) (C) (D)	131 (A) (B) (C) (D)
7 (A) (B) (C) (D)	32 (A) (B) (C) (D)	57 (A) (B) (C) (D)	82 (A) (B) (C) (D)	107 (A) (B) (C) (D)	132 (A) (B) (C) (D)
8 (A) (B) (C) (D)	33 (A) (B) (C) (D)	58 (A) (B) (C) (D)	83 (A) (B) (C) (D)	108 (A) (B) (C) (D)	133 (A) (B) (C) (D)
9 (A) (B) (C) (D)	34 (A) (B) (C) (D)	59 (A) (B) (C) (D)	84 (A) (B) (C) (D)	109 (A) (B) (C) (D)	134 (A) (B) (C) (D)
10 (A) (B) (C) (D)	35 (A) (B) (C) (D)	60 (A) (B) (C) (D)	85 (A) (B) (C) (D)	110 (A) (B) (C) (D)	135 (A) (B) (C) (D)
11 (A) (B) (C) (D)	36 (A) (B) (C) (D)	61 (A) (B) (C) (D)	86 (A) (B) (C) (D)	111 (A) (B) (C) (D)	136 (A) (B) (C) (D)
12 (A) (B) (C) (D)	37 (A) (B) (C) (D)	62 (A) (B) (C) (D)	87 (A) (B) (C) (D)	112 (A) (B) (C) (D)	137 (A) (B) (C) (D)
13 (A) (B) (C) (D)	38 (A) (B) (C) (D)	63 (A) (B) (C) (D)	88 (A) (B) (C) (D)	113 (A) (B) (C) (D)	138 (A) (B) (C) (D)
14 (A) (B) (C) (D)	39 (A) (B) (C) (D)	64 (A) (B) (C) (D)	89 (A) (B) (C) (D)	114 (A) (B) (C) (D)	139 (A) (B) (C) (D)
15 (A) (B) (C) (D)	40 (A) (B) (C) (D)	65 (A) (B) (C) (D)	90 (A) (B) (C) (D)	115 (A) (B) (C) (D)	140 (A) (B) (C) (D)
16 (A) (B) (C) (D)	41 (A) (B) (C) (D)	66 (A) (B) (C) (D)	91 (A) (B) (C) (D)	116 (A) (B) (C) (D)	141 (A) (B) (C) (D)
17 (A) (B) (C) (D)	42 (A) (B) (C) (D)	67 (A) (B) (C) (D)	92 (A) (B) (C) (D)	117 (A) (B) (C) (D)	142 (A) (B) (C) (D)
18 (A) (B) (C) (D)	43 (A) (B) (C) (D)	68 (A) (B) (C) (D)	93 (A) (B) (C) (D)	118 (A) (B) (C) (D)	143 (A) (B) (C) (D)
19 (A) (B) (C) (D)	44 (A) (B) (C) (D)	69 (A) (B) (C) (D)	94 (A) (B) (C) (D)	119 (A) (B) (C) (D)	144 (A) (B) (C) (D)
20 (A) (B) (C) (D)	45 (A) (B) (C) (D)	70 (A) (B) (C) (D)	95 (A) (B) (C) (D)	120 (A) (B) (C) (D)	145 (A) (B) (C) (D)
21 (A) (B) (C) (D)	46 (A) (B) (C) (D)	71 (A) (B) (C) (D)	96 (A) (B) (C) (D)	121 (A) (B) (C) (D)	146 (A) (B) (C) (D)
22 (A) (B) (C) (D)	47 (A) (B) (C) (D)	72 (A) (B) (C) (D)	97 (A) (B) (C) (D)	122 (A) (B) (C) (D)	147 (A) (B) (C) (D)
23 (A) (B) (C) (D)	48 (A) (B) (C) (D)	73 (A) (B) (C) (D)	98 (A) (B) (C) (D)	123 (A) (B) (C) (D)	148 (A) (B) (C) (D)
24 (A) (B) (C) (D)	49 (A) (B) (C) (D)	74 (A) (B) (C) (D)	99 (A) (B) (C) (D)	124 (A) (B) (C) (D)	149 (A) (B) (C) (D)
25 (A) (B) (C) (D)	50 (A) (B) (C) (D)	75 (A) (B) (C) (D)	100 (A) (B) (C) (D)	125 (A) (B) (C) (D)	150 (A) (B) (C) (D)

FIREFIGHTER, F. D.

SAMPLE EXAMINATION I

The time allowed for the entire examination is 3½ hours. In order to create the climate of the test to come, that's precisely the time you should allow yourself. Use a watch and keep a record of your time, especially since you may find it convenient to take the test in several sittings.

1. A problem which is likely to be found in a hotel fire which is not usually found in fires involving other types of residences, such as apartment houses, is

   (A) obstructions in hallways and other passageways
   (B) large numbers of persons in relation to the number of rooms
   (C) delay in the transmission of the fire alarm
   (D) many occupants who don't know the location of exits.

2. On the way to work one morning a firefighter noticed a high tension wire which had blown down and was lying across the sidewalk and into the road. The most appropriate course of action for the firefighter to take would be to

   (A) move the wire to one side by means of a stick or branch of a tree and continue on his way
   (B) continue on his way to his fire house and report the situation to the officer on duty
   (C) call the public utility company from the first public telephone which he passes on his way to the fire house
   (D) stand by the wire to warn away passersby and ask one of them to call the public utility company.

3. Generally, firefighters on prevention inspection duty do not inspect the living quarters of private dwellings unless the occupants agree to the inspection. The best explanation why private dwellings are excluded from compulsory inspections is that

   (A) private dwellings seldom catch fire
   (B) fires in private dwellings are more easily extinguished than other types of fires
   (C) people may resent such inspections as an invasion of privacy
   (D) the monetary value of private dwellings is lower than that of other types of occupancies.

4. "Fire lines" are usually established by the police in order to keep bystanders out of the immediate vicinity while firefighters are fighting the fire. Of the following, the best justification for the establishment of these fire lines is to

(A) prevent theft of property from partially destroyed apartments or stores
(B) prevent interference with operations of the firefighters
(C) give privacy to the victims of the fire
(D) help apprehend arsonists in the crowd.

5. All probationary firefighters are trained in both engine company and ladder company operations. The most important reason for training in both types of operations is that

(A) at any fire it may be necessary for engine company people to perform ladder company duties and vice versa
(B) engine company people can better perform their own functions if they are also trained in ladder company functions and vice versa.
(C) after such training, the probationary firefighter can better decide which type of company he or she prefers
(D) after such training, the Fire Department can better decide which people are best suited for each type of company.

6. While on inspection duty, a firefighter discovers the superintendent of a tenement just starting to remove boxes and other material which are blocking hallways. Apparently the superintendent started removal as soon as he saw the firefighter approach. In this situation, it is most important that the firefighter

(A) warn the superintendent of the penalties for violation of the Fire Prevention Code
(B) help the superintendent remove the material blocking the hallways
(C) commend the superintendent for his efforts to maintain a safe building
(D) check again, after completing the inspection, to see whether the material has been removed completely.

7. Loss of an official department badge by a firefighter is a serious offense for which disciplinary action may be taken against the member. Of the following, the main reason that the Fire Department regards this loss so seriously is that

(A) badges are expensive and difficult to replace

(B) loss of a badge usually is an indication of poor morale
(C) the person finding the badge might use it for improper purposes
(D) a firefighter must at all times wear proper identification.

8. At 2 A.M. a passerby observes a fire in an office building. He immediately starts to run to a nearby corner in order to look for a fire alarm box. After running a few steps, he notices a police telephone box on a lamp post and uses it to report the fire. This action is

(A) good chiefly because the police as well as the Fire Department respond to fire alarms
(B) bad chiefly because police telephone boxes are not to be used by civilians under any circumstances
(C) good chiefly because the Fire Department will then receive prompt notification of the fire
(D) bad chiefly because police telephone boxes may be used by the public only for police emergencies.

9. A fire insurance inspector suggested to the manager of a fireproof warehouse that bags of flour be stacked on skids (wooden platforms 6" high, 6x6 feet in area). Of the following, the best justification for this suggestion is that in the event of a fire, the bags on skids are less likely

(A) to topple
(B) to be damaged by water used in extinguishment
(C) to catch fire
(D) to be ripped by fire equipment.

10. Permitting piles of scrap paper cuttings to accumulate in a factory building is a bad practice chiefly because they may

(A) ignite spontaneously
(B) interfere with fire extinguishment operations
(C) catch fire from a spark or smoldering match
(D) interfere with escape of occupants if a fire occurs.

11. High grass and weeds should not be permitted to grow near a building chiefly because, in the event of a grass fire, the weeds and grass may

(A) give off toxic fumes
(B) limit maneuverability of firefighters
(C) interfere with the escape of occupants from the building
(D) bring the fire to the building and set it on fire.

12. Visitors near patients in "oxygen tents" are not permitted to smoke. The best of the following reasons for this prohibition is that

(A) the flame of the cigarette or cigar may flare dangerously
(B) smoking tobacco is irritating to persons with respiratory diseases
(C) smoking in bed is one of the major causes of fires
(D) diseases may be transmitted by means of tobacco smoke.

13. At a hot and smoky fire, a lieutenant ordered firefighters to work in pairs. Of the following, the best justification of this order is that

(A) better communications result since one person can bring messages back to the lieutenant
(B) more efficient operation results since many vital activities require two people
(C) better morale results since firefighters are more willing to face danger in pairs
(D) safer operations result since one firefighter can help the other, if disabled.

14. Firefighters frequently open windows, doors, and skylights of a building on fire in a planned or systematic way in order to ventilate the building. What is least likely to be accomplished by ventilation is the

(A) increase in visibility of firefighters
(B) slowdown in the rate of burning of the building
(C) reduction in the danger from toxic gases
(D) control of the direction of travel of the fire.

15. At the first sign of a fire, the manager of a motion picture theatre had the lights turned on and made the following announcement: "Ladies and gentlemen, the management has found it necessary to dismiss the audience. Please remain seated until it is time for your aisle to file out. In leaving the theatre follow the directions of the ushers. There is no danger involved." The manager's action in this situation was

(A) proper
(B) improper chiefly because he did not tell the audience the reason for the dismissal
(C) improper chiefly because he did not permit all members of the audience to leave at once
(D) improper chiefly because he misled the audience by saying that there was no danger.

16. Generally, sprinkler heads must be replaced each time they are used. The best explanation of why this is necessary is that the sprinkler heads

(A) are subject to rusting after discharging water
(B) may become clogged after discharging water
(C) have a distorted pattern of discharge of water after use
(D) are set off by the effect of heat on metal and cannot be re-set.

17. After a fire in an apartment had been brought under control, firefighters were engaged in extinguishing the last traces of fire. A firefighter who noticed an expensive vase in the room in which activities were concentrated moved it to an empty closet in another room. The firefighter's action was

(A) proper chiefly because the owners would realize that extreme care was taken to avoid damage to their possessions
(B) improper chiefly because disturbance of personal possessions should be kept to a minimum
(C) proper chiefly because the chance of damage is reduced
(D) improper chiefly because the firefighter should have devoted this time to putting out the fire.

18. While standing in front of a fire house a firefighter is approached by a woman with a baby carriage. The woman asks the firefighter if he will "keep an eye" on the baby while she visits a doctor in a nearby building. In this situation, the best course of action for the firefighter is to

(A) agree to the woman's request but warn her that it may be necessary to leave to answer an alarm
(B) refuse politely after explaining that he or she is on duty and may not become involved in other activities
(C) refer the woman to the officer on duty
(D) ask the officer on duty for permission to grant the favor.

19. Suppose that, while cooking, a pan of grease catches fire. The method which would be most effective, if available, in putting out the fire is to

(A) dash a bucket of water on the fire
(B) direct a stream of water on the fire from a fire extinguisher
(C) pour a bottle of household ammonia water on the fire
(D) empty a box of baking soda on the fire.

20. Persons engaged in certain hazardous activities are required to obtain a Fire Department permit or certificate for which a fee is charged. The main reason for requiring permits or certificates is to

(A) obtain revenue for the city government
(B) prevent unqualified persons from engaging in these activities
(C) obtain information about these activities in order to plan for fire emergencies
(D) warn the public of the hazardous nature of these activities.

21. A firefighter, on his way to work, is stopped by a citizen who complains that the employees of a nearby store frequently pile empty crates and boxes in a doorway, blocking passage. The most appropriate action for the firefighter to take is to

(A) assure the citizen that the fire department's inspectional activities will eventually "catch up" with the store
(B) obtain the address of the store and investigate to determine whether the citizen's complaint is justified
(C) obtain the address of the store and report the complaint to his or her superior officer
(D) ask the citizen for specific dates on which this practice has occurred to determine whether the complaint is justified.

22. The crime of arson is defined as the malicious burning of a house or other property. The best illustration of arson is a fire

(A) in an apartment started by a four year old boy playing with matches
(B) in a barn started by a drunken man who overturns a lantern
(C) in a store started by the bankrupt owner in order to collect insurance
(D) in a house started by a neighbor who carelessly burned leaves in his garden.

23. The Fire Department now uses companies on fire duty, with their apparatus, for fire prevention inspection in commercial buildings. The change which was most important in making this inspection procedure practicable was the

(A) reduction of hours of work of firefighters
(B) use of two-way radio equipment

(C) use of enclosed cabs on fire apparatus
(D) increase in property values during the post-war period.

24. Many fires are caused by improper use of oxy-acetylene torches. The main cause of such fires is the

(A) high pressure under which the gases are stored
(B) failure to control or extinguish sparks
(C) high temperatures generated by the equipment
(D) explosive nature of the gases used.

25. Members of the Fire Department are permitted to engage in outside employment during their off-duty time under certain conditions. However, members are not permitted to work as carpenters specializing in the elimination of violations of the Fire Prevention Code because such work

(A) requires great skill and knowledge
(B) might be inspected by them or other firefighters
(C) may cause injury to the firefighters
(D) would deprive a worker of the means of earning a living.

26. The most important reason for having members of the Fire Department wear uniforms is to

(A) indicate the semi-military nature of the Fire Department
(B) build morale and esprit de corps of members
(C) identify members on duty to the public and other members
(D) provide clothing suitable for the work performed.

27. Of the following types of fires, the one which is likely to have the <u>least</u> amount of damage from water used in extinguishment is a fire in

(A) a rubber toy factory
(B) a retail hardware store
(C) an outdoor lumber yard
(D) a furniture warehouse.

28. When fighting fires involving unevenly piled goods, it is particularly important that the water streams penetrate all parts of the goods exposed to the fire. The position of hose nozzle which will provide maximum water penetration is

(A) above the fire
(B) to the side away from the wind
(C) on level with the fire
(D) to the side facing the wind.

29. "In case of suspected arson, it is important that firefighters engaged in fighting the fire remember conditions that existed at the time of their arrival. Particular attention should be given to all doors and windows." The main justification for this statement is that knowledge of the condition of the doors and windows may indicate

(A) where the fire started
(B) who set the fire
(C) the best way to ventilate the building
(D) that someone wanted to prevent extinguishment.

30. While visiting a hotel, a firefighter discovers a fire which apparently has been burning for some time and is rapidly spreading. Of the following, the first action for the firefighter to take is to

(A) find the nearest fire extinguisher and attempt to put out the fire
(B) notify the desk clerk of the fire
(C) send an alarm from the nearest street alarm box
(D) run throughout the hotel and warn all occupants to evacuate the building.

31. A firefighter inspecting buildings in a commercial area came to one whose outside surface appeared to be of natural stone. The owner told the firefighter that it was not necessary to inspect his building as it was "fireproof." The firefighter, however, completed inspection of the building. Of the following, the best reason for continuing the inspection is that

(A) stone buildings catch fire as readily as wooden buildings
(B) the Fire Department cannot make exceptions in its inspection procedures
(C) the building may have been built of imitation stone
(D) interiors and contents of stone buildings can catch fire.

32. That which is <u>least</u> valid as a reason for the Fire Department to investigate the causes of fire is to

(A) determine whether the fire was the result of arson
(B) estimate the amount of loss for insurance purposes
(C) gather information useful in fire prevention
(D) discover violations of the Fire Prevention Code.

33. While on duty at a fire, a probationary firefighter receives an order from the lieutenant which appears to conflict with the principles of firefighting taught at the fire school. Of the following, the best course of action for the firefighter to take is to follow the order and at a convenient time, after the fire, to

(A) discuss the apparent inconsistency with the lieutenant
(B) discuss the apparent inconsistency with another officer
(C) mention this apparent inconsistency in an informal discussion
(D) ask a more experienced firefighter about the apparent inconsistency.

34. When fighting fires on piers, the Fire Department frequently drafts salt water from the harbor. The chief advantage of using harbor water instead of relying on water from street mains is that harbor water is

(A) less likely to cause water damage
(B) available in unlimited quantities
(C) more effective in extinguishing fires due to its salt content
(D) less likely to freeze in low temperatures due to its salt content

35. Firefighters always try to keep to a minimum the amount of water used in extinguishing a fire without reducing the effectiveness of their operation. Of the following reasons for New York City firefighters using water sparingly, the <u>least</u> valid is that the use of excess water may

(A) dangerously overload a building and cause its collapse
(B) flood the subway system or damage other public utilities
(C) damage the contents of the building on fire
(D) dangerously reduce the water supply of the city.

Questions 36 to 39 inclusive are based upon the following paragraph:

The canister type gas mask consists of a tight-fitting face piece connected to a canister containing chemicals which filter toxic gases and smoke from otherwise breathable air. These masks are of value when used with due regard to the fact that two or three percent of gas in air is about the highest concentration that the chemicals in the canister will absorb and that these masks do not provide the oxygen which is necessary for the support of life. In general, if flame is visible, there is sufficient oxygen for firefighters although toxic gases may be present. Where there is heavy smoke and no flame, an oxygen deficiency may exist. Fatalities have occurred where filter type canister masks have been used in attempting rescue from manholes, wells, basements, or other locations deficient in oxygen.

36. If the mask described above is used in an atmosphere containing oxygen, nitrogen, and carbon monoxide, we would expect the mask to remove from the air breathed

(A) the nitrogen only
(B) the carbon monoxide only
(C) the nitrogen and the carbon monoxide
(D) none of these gases.

37. According to this paragraph, when firefighters are wearing these masks at a fire where flame is visible, the firefighters can generally feel that as far as breathing is concerned, they are

(A) safe, since the mask will provide them with sufficient oxygen to live
(B) unsafe, unless the gas concentration is below 2 or 3 percent
(C) safe, provided the gas concentration is above 2 or 3 percent
(D) unsafe, since the mask will not provide them with sufficient oxygen to live.

38. According to this paragraph, fatalities have occurred to persons using this type gas mask in manholes, wells, and basements because

(A) the supply of oxygen provided by the mask ran out
(B) the air in those places did not contain enough oxygen to support life
(C) heavy smoke interfered with the operation of the mask
(D) the chemicals in the canister did not function properly.

39. The following short-hand formula may be used to show, in general, the operation of the gas mask described in the preceding paragraph:

(Chemicals in canister)⟶(Air + gases) = Breathable Air.

The arrow in the formula, when expressed in words, means most nearly

(A) replace
(B) are changed into
(C) act upon
(D) give off.

Questions 40 to 42 inclusive are based upon the following paragraph:

The only openings permitted in fire partitions except openings for ventilating ducts shall be those required for doors. There shall be but one such door opening unless the provision of additional openings would not exceed in total width of all doorways 25 percent of the length of the wall. The minimum distance between openings shall be three feet. The maximum area for such a door opening shall be 80 square feet, except that such openings for the passage of motor trucks may be a maximum of 140 square feet.

40. According to the above paragraph, openings in fire partitions are permitted only for

(A) doors
(B) doors and windows
(C) doors and ventilation ducts
(D) doors, windows, and ventilation ducts.

41. In a fire partition 22 feet long and 10 feet high, the maximum number of doors 3 feet wide and 7 feet high is

(A) 1
(B) 2
(C) 3
(D) 4.

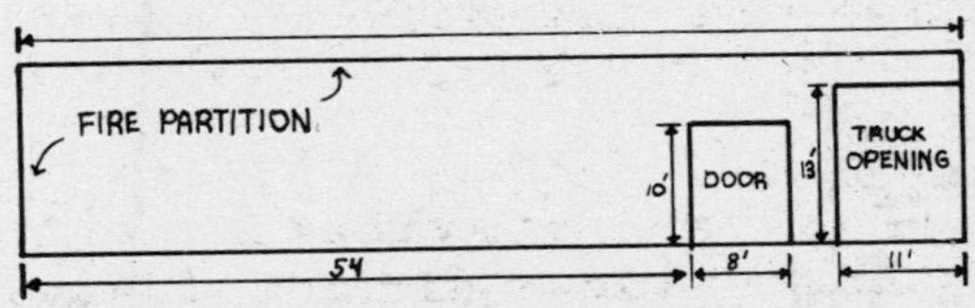

42. The most accurate statement about the layout shown above is that the

(A) total width of the openings is too large
(B) truck opening is too large
(C) truck and door openings are too close together
(D) layout is acceptable.

43. At a given temperature a wet hand will freeze to a bar of metal, but not to a piece of wood, because

(A) the metal expands and contracts more than the wood
(B) the wood is softer than the metal
(C) the wood will burn at a lower temperature than the metal
(D) the metal is a better conductor of heat than the wood.

44. Of the following items commonly found in a household, the one that uses the most electric current is

(A) a 150 watt light bulb
(B) a toaster
(C) a door buzzer
(D) an 8" electric fan.

Questions 45 to 47 inclusive are based upon the following paragraph:

The average daily flow of water through public water systems in American cities ranges generally between 40 and 250 gallons per capita, depending upon the underground leakage in the system, the amount of waste in domestic premises, and the quantity used for industrial purposes. The problem of supplying this water has become serious in many cities. Supplies, once adequate, in many cases have become seriously deficient, due to greater demands with increased population, and growing industrial use of water. Water works, operating on fixed schedules of water charges, have in many cases not been able to afford the heavy capital expenditures necessary to provide adequate supply, storage and distribution facilities. Thus the adequacy of a public water supply for fire protection in any given location cannot properly be taken for granted.

45. The four programs listed below are possible ways by which American communities might try to reduce the seriousness of the water shortage problem. The one program which does not directly follow from the paragraph above is that of

(A) regular replacement of old street water mains by new ones
(B) inspection and repair of leaky plumbing fixtures
(C) fire prevention inspection and education to reduce the amount of water used to extinguish fires
(D) research into industrial processes to reduce the amount of water used in those processes.

46. The main conclusion reached by the author of the above paragraph is

(A) there is a waste of precious natural resources in America
(B) communities have failed to control the industrial use of water
(C) a need exists for increasing the revenue of water works to build up adequate supplies of water
(D) fire departments cannot assume that they will always have the necessary supply of water available to fight fires.

47. Per capita consumption of water of a community is determined by the formula

(A) $\frac{\text{Population}}{\text{Total consumption in gallons}} =$

Per capita consumption in gallons

(B) $\frac{\text{Total consumption in gallons}}{\text{Population}} =$

Per capita consumption in gallons

(C) Total consumption in gallons X Population = Per capita consumption in gallons

(D) Total consumption in gallons - Population = Per capita consumption in gallons.

Questions 48 to 51 inclusive are based upon the following paragraph:

An annual leave allowance, which combines leaves previously given for vacation, personal business, family illness and other reasons shall be granted members. Calculation of credits for such leave shall be on an annual basis beginning January 1st of each year. Annual leave credits shall be based on time served by members during preceding calendar year. However, when credits have been accrued and member retires during current year, additional annual leave credits shall, in this instance, be granted at accrual rate of three days for each completed month of service, excluding terminal leave. If accruals granted for completed months of service extend into following month, member shall be granted an additional three days accrual for completed month. This shall be the only condition where accruals in a current year are granted for vacation period in such year.

48. According to the above paragraph, if a firefighter's spouse were to become seriously ill so that it was necessary to take time off from work to be with her (or him), such time off would be deducted from the firefighter's

(A) annual leave allowance
(B) vacation leave allowance
(C) personal business leave allowance
(D) family illness leave allowance.

49. Terminal leave means leave taken

(A) at the end of the calendar year
(B) at the end of the vacation year
(C) immediately before retirement
(D) before actually earned, because of an emergency.

50. A firefighter appointed on July 1, 1980 will be able to take his first full or normal annual leave during the period

(A) July 1, 1980 to June 30, 1981
(B) Jan. 1, 1981 to Dec. 31, 1981
(C) July 1, 1981 to June 30, 1982
(D) Jan. 1, 1982 to Dec. 31, 1982.

51. According to this paragraph, a member who retires on July 15 of this year will be entitled to receive leave allowance, based on this year, of

(A) 15 days
(B) 18 days
(C) 21 days
(D) 24 days.

Questions 52 to 55 inclusive are based upon the following paragraph:

During fire operations all members shall be constantly alert to the possibility of the crime of arson. In the event conditions indicate this possibility, the officer in command shall promptly notify the Fire Marshal. Unauthorized persons shall be prohibited from entering premises and actions of those authorized carefully noted. Members shall refrain from discussion of the fire and prevent disturbance of essential evidence. If necessary, the officer in command shall detail one or other members at location with information for the Fire Marshal upon his or her arrival.

52. From the above paragraph, it may be inferred that the reason for prohibiting unauthorized persons from entering the fire premises when arson is suspected is to prevent such persons from

(A) endangering themselves in the fire
(B) interfering with the firefighters fighting the fire
(C) disturbing any evidence of arson
(D) committing acts of arson.

53. The title which best describes the subject matter of the above paragraph is

(A) Techniques of Arson Detection
(B) The Role of the Fire Marshal in Arson Cases
(C) Fire Scene Procedures in Cases of Suspected Arson
(D) Evidence in Arson Investigations.

54. The statement that is most correct and complete is that the responsibility for detecting signs of arson at a fire belongs to the

(A) Fire Marshal
(B) Fire Marshal and officer in command

(C) Fire Marshal, officer in command, and any members detailed at location with information for the Fire Marshal
(D) members present at the scene of the fire regardless of their rank or position.

55. From the above paragraph it may be inferred that the Fire Marshal usually arrives at the scene of a fire

(A) before the fire companies
(B) simultaneously with the fire companies
(C) immediately after the fire companies
(D) some time after the fire companies.

56. Sand and ashes are frequently placed on icy pavements to prevent skidding. The effect of the sand and ashes is to increase

(A) inertia
(B) gravity
(C) momentum
(D) friction.

57. The air near the ceiling of a room is usually warmer than the air near the floor because

(A) there is better air circulation at floor level
(B) warm air is lighter than cold air
(C) windows are usually nearer the floor than the ceiling
(D) heating pipes usually run along the ceiling.

58. It is safer to use the ladder positioned as shown in diagram 1 than as shown in diagram 2 because in diagram 1

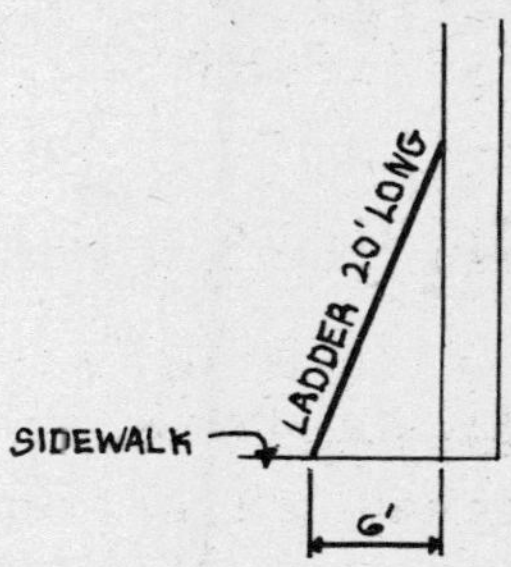

DIA. 1

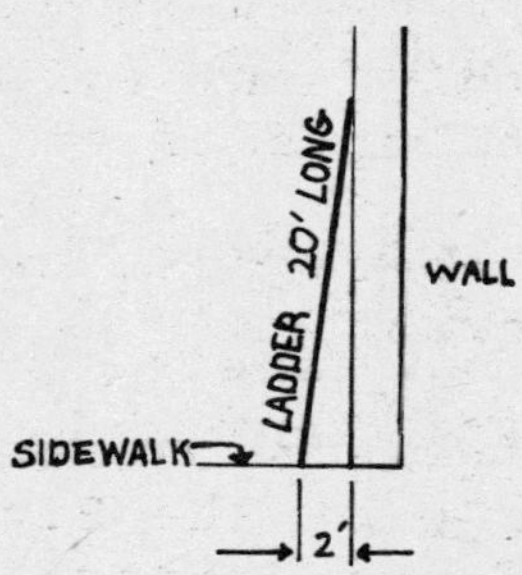

DIA. 2

(A) less strain is placed upon the center rungs of the ladder
(B) it is easier to grip and stand on the ladder
(C) the ladder reaches a lower height
(D) the ladder is less likely to tip over backwards.

59. It is easier to pour a liquid from

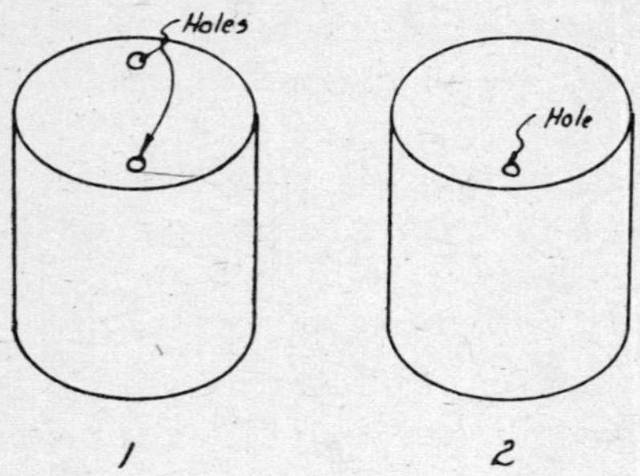

(A) Can 1 because there are two holes from which the liquid can flow
(B) Can 1 because air can enter through one hole while the liquid comes out of the other hole
(C) Can 2 because the liquid comes out under greater pressure
(D) Can 2 because it is easier to direct the flow of the liquid when there is only one hole.

60. A substance which is subject to spontaneous combustion is one that

(A) is explosive when heated
(B) is capable of catching fire without an external source of heat
(C) acts to speed up the burning of material
(D) liberates oxygen when heated.

61. The sudden shutting down of a nozzle on a hose discharging water under high pressure is a bad practice chiefly because

(A) the hose is likely to whip about violently
(B) the hose is likely to burst
(C) the valve handle is likely to snap
(D) the valve handle is likely to jam.

62. Fire can continue where there are present fuel, oxygen from the air or other source, and a sufficiently high temperature to maintain combustion. The method of extinguishment of fire <u>most commonly</u> used is to

(A) remove the fuel
(B) exclude the oxygen from the burning material
(C) reduce the temperature of the burning material
(D) smother the flames of the burning material.

63. The siphon arrangement shown below which would most quickly transfer a solution from the container on the left side to the one on the right side is numbered

(A) 1 (C) 3
(B) 2 (D) 4.

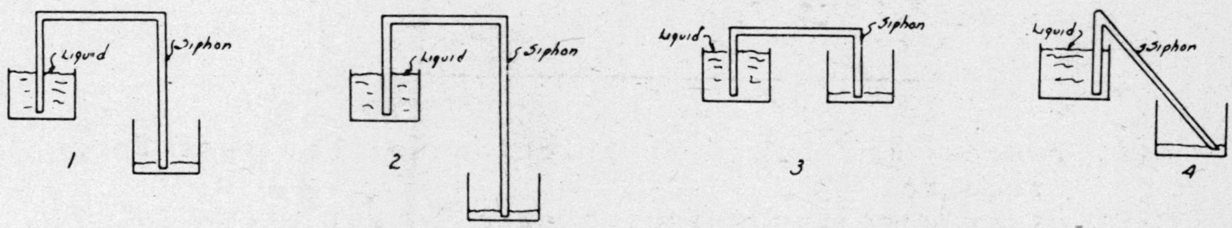

64. Static electricity is a hazard in industry chiefly because it may cause

(A) dangerous or painful burns
(B) chemical decomposition of toxic elements
(C) sparks which can start an explosion
(D) overheating of electrical equipment.

65. The strongest method of supporting the shelf is shown in diagram

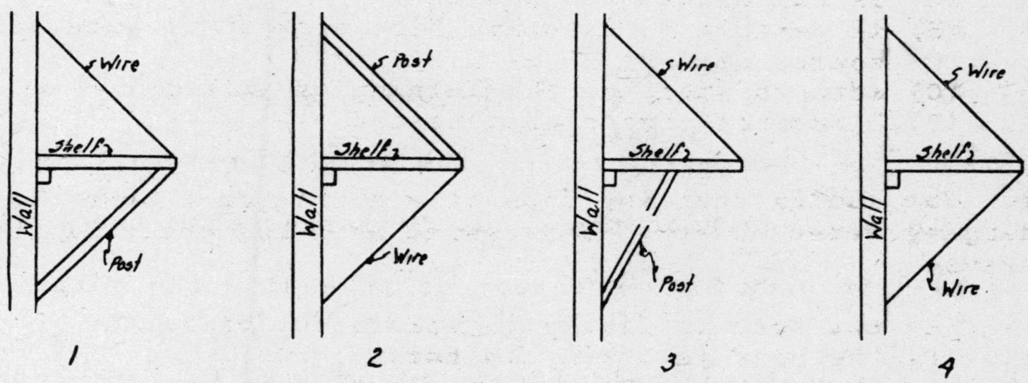

(A) 1 (C) 3
(B) 2 (D) 4.

66. A row boat will float deeper in fresh water than in salt water because

(A) in the salt water, the salt will occupy part of the space
(B) fresh water is heavier than salt water
(C) salt water is heavier than fresh water
(D) salt water offers less resistance than fresh water.

67. It is easier to get the load onto the platform by using the ramp than it is to lift it directly onto the platform. This is true because the effect of the ramp is to

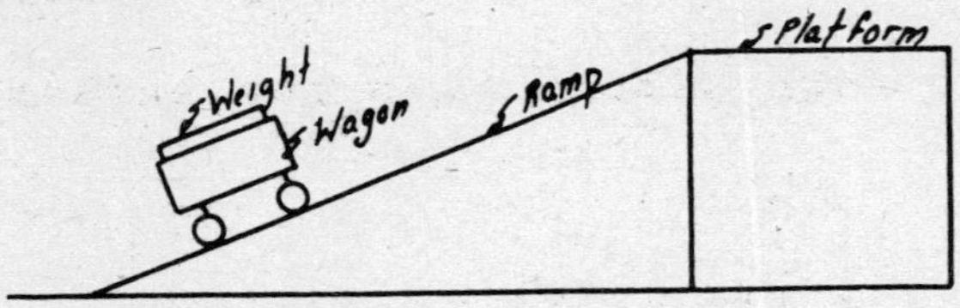

(A) reduce the amount of friction so that less force is required
(B) distribute the weight over a larger area
(C) support part of the load so that less force is needed to move the wagon
(D) increase the effect of the moving weight.

68. More weight can be lifted by the method shown in diagram 2 than as shown in diagram 1 because

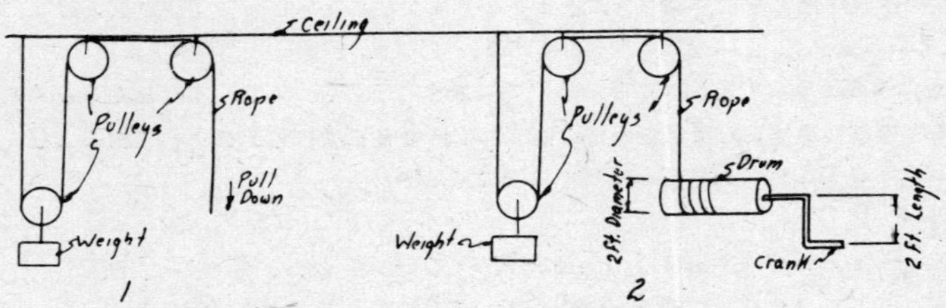

(A) it takes less force to turn a crank than it does to pull in a straight line
(B) the drum will prevent the weight from falling by itself
(C) the length of the crank is larger than the radius of the drum
(D) the drum has more rope on it easing the pull.

69. As the endless chain is pulled down in the direction shown, the weight

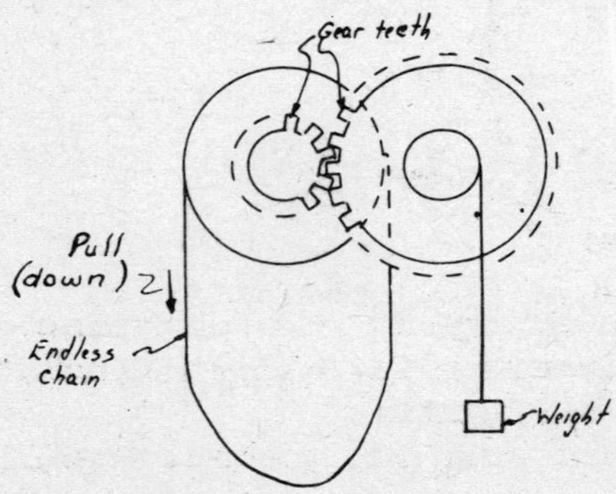

(A) will move up faster than the endless chain is pulled down
(B) will move up slower than the endless chain is pulled down
(C) will move down faster than the endless chain is pulled down
(D) will move down slower than the endless chain is pulled down.

70. Two balls of the same size, but different weights, are both dropped from a 10-ft. height. The statement that is most accurate is that

(A) both balls will reach the ground at the same time because they are the same size
(B) both balls will reach the ground at the same time because the effect of gravity is the same on both balls
(C) the heavier ball will reach the ground first because it weighs more
(D) the lighter ball will reach the ground first because air resistance is greater on the heavier ball.

71. It is considered poor practice to increase the leverage of a wrench by placing a pipe over the handle of the wrench. This is true principally because

(A) the wrench is likely to break
(B) the wrench may slip off the nut
(C) it is harder to place the wrench on the nut
(D) the wrench is more difficult to handle.

72. If wheel #1 is turned in the direction shown, wheel #2 will

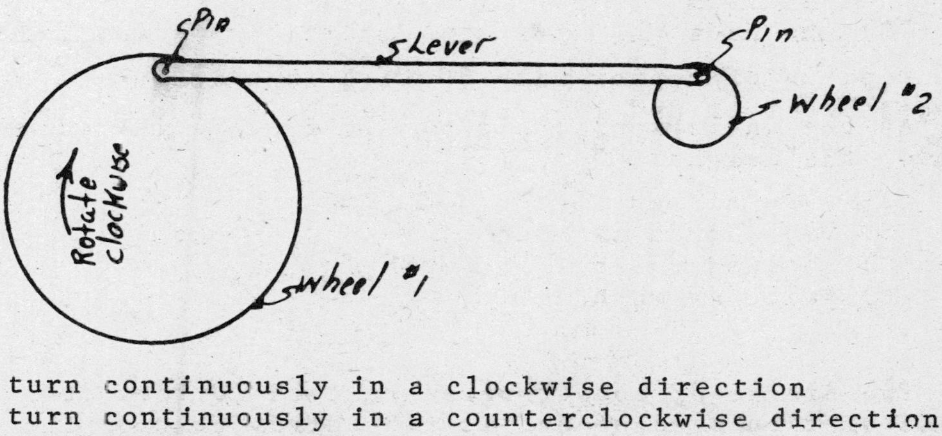

(A) turn continuously in a clockwise direction
(B) turn continuously in a counterclockwise direction
(C) move back and forth
(D) become jammed and both wheels will stop.

73. All solid areas represent equal weights attached to the flywheel.

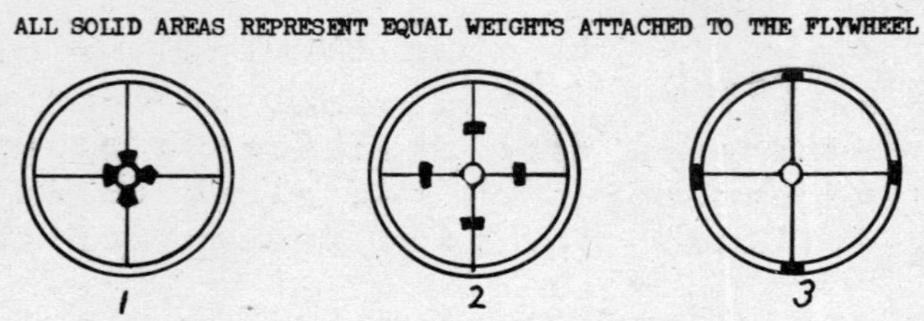

The above diagrams are of flywheels made of the same material with the same dimensions and attached to similar engines. The solid areas represent equal weights attached to the flywheel. If all three engines are running at the same speed for the same length of time, and the power to the engines is shut off simultaneously

(A) wheel 1 will continue turning longest
(B) wheel 2 will continue turning longest
(C) wheel 3 will continue turning longest
(D) all three wheels will continue turning for the same time.

74. The substance which expands when freezing is

(A) alcohol
(B) ammonia
(C) mercury
(D) water.

Questions 75 and 76 are to be answered in accordance with the information in the following statement:

The electrical resistance of copper wires varies directly with their lengths and inversely with their cross section areas.

75. A piece of copper wire 30 feet long is cut into two pieces, 20 feet and 10 feet. The resistance of the longer piece, compared to the shorter, is

(A) one-half as much
(B) two-thirds as much
(C) one and one-half as much
(D) twice as much.

76. Two pieces of copper wire are each 10 feet long but the cross section area of one is two-thirds that of the other. The resistance of the piece with the larger cross section area is

(A) one-half the resistance of the smaller
(B) two-thirds the resistance of the smaller
(C) one and one-half times the resistance of the smaller
(D) twice the resistance of the smaller.

77. When drilling a hole through a piece of wood with an auger bit, it is considered good practice to clamp a piece of scrap wood to the underside of the piece through which the hole is being drilled. The main reason for this is to

(A) direct the auger bit
(B) speed up the drilling operation
(C) prevent the drill from wobbling
(D) prevent splintering of the wood.

78. The arrangement of the lever which would require the least amount of force to move the weight is shown in

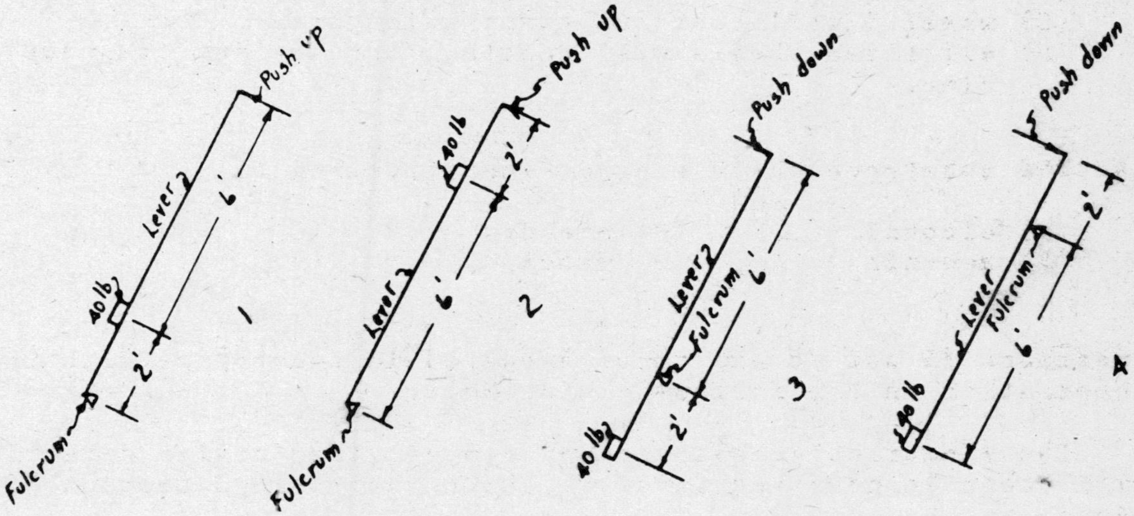

(A) 1
(B) 2
(C) 3
(D) 4.

79. Steel supporting beams in buildings are often surrounded by a thin layer of concrete to keep the beams from becoming hot and collapsing during a fire. The statement which best explains how collapse is prevented by this arrangement is that concrete

(A) becomes stronger as its temperature is increased
(B) acts as an insulating material
(C) protects the beam from rust and corrosion
(D) reacts chemically with steel at high temperatures.

80. If boiling water is poured into a drinking glass, the glass is likely to crack. If, however, a metal spoon is first placed in the glass, it is much less likely to crack. The reason that the glass with the spoon is less likely to crack is that the spoon

(A) distributes the water over a larger surface of the glass
(B) quickly absorbs heat from the water
(C) reinforces the glass
(D) reduces the amount of water which can be poured into the glass

81. It takes more energy to force water through a long pipe than through a short pipe of the same diameter. The principal reason for this is

(A) gravity
(B) friction
(C) inertia
(D) cohesion.

82. A pump discharging at 300 lbs. per sq. inch pressure delivers water through 100 feet of pipe laid horizontally. If the valve at the end of the pipe is shut so that no water can flow, then the pressure at the valve is, for practical purposes

(A) greater than the pressure at the pump
(B) equal to the pressure at the pump
(C) less than the pressure at the pump
(D) greater or less than the pressure at the pump, depending on the type of pump used.

83. The explosive force of a gas when stored under various pressures is given in the following table:

Storage Pressure	Explosive Force
10	1
20	8
30	27
40	64
50	125

The statement which best expresses the relationship between storage pressure and explosive force is that

(A) there is no systematic relationship between an increase in storage pressure and an increase in explosive force
(B) the explosive force varies as the square of the pressure
(C) the explosive force varies as the cube of the pressure
(D) the explosive force varies as the fourth power of the pressure.

84. The leverage system in the sketch above is used to raise a weight. In order to reduce the amount of force required to raise the weight, it is necessary to

(A) decrease the length of the lever
(B) place the weight closer to the fulcrum
(C) move the weight closer to the person applying the force
(D) move the fulcrum farther from the weight.

The following four questions, numbered 85 to 88, relate to the information given in the following paragraph and are to be answered in accordance with this information.

Old buildings are liable to possess a special degree of fire risk merely because they are old. Outmoded electrical wiring systems and installation of new heating appliances for which the building was not designed may contribute to the increased hazard. Old buildings have often been altered many times; parts of the structure may antedate building codes; dangerous defects may have been covered up. On the average, old buildings contain more lumber than comparable new buildings which, in itself, makes old buildings more susceptible to fire. It is not true, though, that sound lumber in old buildings is drier than new lumber. Moisture content of lumber varies with that of the atmosphere to which it is exposed.

85. According to the above paragraph, old buildings present a special fire hazard chiefly because of the

(A) poor planning of the buildings when first designed
(B) haphazard alteration of the buildings
(C) failure to replace worn out equipment
(D) inadequate enforcement of the building codes.

86. We may conclude from the above paragraph that lumber

(A) should not be used in buildings unless absolutely necessary
(B) should not be used near electrical equipment
(C) is more inflammable than newer types of building materials
(D) tends to lose its moisture at a constant rate.

87. According to the above paragraph, the amount of moisture in the wooden parts of a building depends upon the

(A) age of the building
(B) moisture in the surrounding air
(C) type of heating equipment used in the building
(D) quality of lumber used.

88. In regard to building codes, the above paragraph implies that

(A) old buildings are exempt from the provisions of building codes
(B) some buildings now in use were built before building codes were adopted
(C) building codes usually don't cover electrical wiring systems
(D) building codes are generally inadequate.

89. An officer must familiarize him or herself with first aid methods. Of the following, the best reason for the above rule is

(A) it saves expense to have someone take care of simple injuries not requiring a doctor
(B) it may keep the Authority out of a law suit if a simple injury is taken care of promptly
(C) it is of great advantage that an officer be prepared to act quickly in an emergency
(D) in doing first aid, it is sometimes better to do nothing than to do the wrong thing.

90. The method which is not a recommended procedure for treating shock is

(A) administering stimulants
(B) placing in a horizontal position
(C) applying heat
(D) applying direct pressure.

91. A combination of certain symptoms, following an atomic bomb attack, would be strongly indicative that a man or woman was suffering from radiation sickness. Which one of the following, in such a situation, would be <u>least</u> likely to be a part of that combination of symptoms?

(A) frequent loose bowel movements
(B) a feeling of pressure within the chest
(C) sore throat
(D) vomiting.

92. Never forget to loosen a tourniquet every once in a while and let the blood flow. Of the following, the best justification for this statement is that

(A) a tourniquet is merely an emergency measure
(B) a tourniquet tends to lose its effectiveness unless it is tightened
(C) heavy loss of blood may weaken the victim seriously
(D) prolonged interruption of blood circulation may cause gangrene.

93. By a compound fracture is generally meant a fracture in which

(A) the broken bone has penetrated the skin
(B) the bone is broken at more than one point
(C) the bone has not only been broken but has been dislocated as well
(D) more than one bone has been broken.

94. Studies have indicated that there is a correlation between an individual's physical well-being and his susceptibility to accidents. Since most of the duties of firefighters involve an element of danger, it therefore follows that

(A) firefighters should be asked how they are feeling before being assigned to a hazardous task by their superior officer
(B) all firefighters should have the privilege of unlimited paid sick leave
(C) a claim of illness should automatically excuse a firefighter from all duty
(D) a firefighter who does not feel "up to snuff" should report in ill.

95. The newly appointed firefighter should realize that the principal function of an immediate superior officer is to

(A) inflict discipline for rule infractions
(B) direct the activities of subordinates so that they function in a proper acceptable manner
(C) make recommendations to promote the most able subordinates under his or her command
(D) keep records that properly reflect the activities of his subordinates.

96. Firefighters who are eager for promotion and who thoroughly prepare themselves for promotion examinations are likely to become better firefighters. The most probable reason for this is

(A) only the better firefighters want to be promoted
(B) firefighters who do not seek promotion are convinced of their limitations
(C) the study material necessary for promotion preparation is mainly concerned with the duties of the position
(D) firefighters who spend their free time preparing for promotion are likely to stay out of trouble.

97. A newly appointed firefighter should be aware that the main purpose of constructive criticism made by a superior officer would be to

(A) improve his or her performance
(B) enhance his or her opportunities for promotion
(C) display authority
(D) encourage the firefighter to participate in company social activities.

98. A newly appointed firefighter should be cognizant of the fact that a large part of each day will be spent on duty in the firehouse waiting for alarms to be sounded. Although some time will be devoted to the maintenance of equipment, there will still be a considerable amount of idle time. It would therefore be most appropriate to arrive for the first day's duty

(A) with extensive reading material
(B) with the tools of a hobby
(C) without preconceived ideas about how to spend spare time
(D) enrolled in a home study course.

99. During the course of fighting a fire a superior officer notices a subordinate performing an important function in an improper manner. It would be most appropriate for the superior officer to correct the firefighter

(A) after the fire has been extinguished, in the privacy of the firehouse
(B) immediately
(C) immediately only if the public is not in attendance and the firefighter would not be embarrassed
(D) at the next regularly scheduled training session.

100. A firefighter should be aware that the main purpose of disciplinary action taken by superior officers in a fire department against their subordinates is to

(A) inflict punishment for rule infractions
(B) train subordinates to perform their duties properly
(C) show "who is boss"
(D) establish rules of acceptable behavior.

SAMPLE EXAMINATION I -ANSWERS

1.	D	26.	C	51.	C	76.	B
2.	D	27.	C	52.	C	77.	D
3.	C	28.	A	53.	C	78.	A
4.	B	29.	D	54.	D	79.	B
5.	A	30.	C	55.	D	80.	B
6.	D	31.	D	56.	D	81.	B
7.	C	32.	B	57.	B	82.	B
8.	C	33.	A	58.	D	83.	C
9.	B	34.	B	59.	B	84.	B
10.	C	35.	D	60.	B	85.	B
11.	D	36.	B	61.	B	86.	C
12.	A	37.	B	62.	C	87.	B
13.	D	38.	B	63.	B	88.	B
14.	B	39.	C	64.	C	89.	C
15.	A	40.	C	65.	A	90.	D
16.	D	41.	A	66.	C	91.	B
17.	C	42.	B	67.	C	92.	D
18.	B	43.	D	68.	C	93.	A
19.	D	44.	B	69.	D	94.	D
20.	B	45.	C	70.	B	95.	B
21.	C	46.	D	71.	B	96.	C
22.	C	47.	B	72.	D	97.	A
23.	B	48.	A	73.	C	98.	C
24.	B	49.	C	74.	D	99.	B
25.	B	50.	D	75.	D	100.	B

FIREFIGHTER, F. D.

SAMPLE EXAMINATION II

The time allowed for the entire examination is 3½ hours. In order to create the climate of the test to come, that's precisely the time you should allow yourself. Use a watch and keep a record of your time, especially since you may find it convenient to take the test in several sittings.

1. Each year many children die in fires which they have started while playing with matches. Of the following measures, the one that would be most effective in preventing such tragedies is to

(A) warn the children of the dangers involved
(B) punish parents who are found guilty of neglecting their children
(C) keep matches out of the reach of children
(D) use only safety matches.

2. Holding a fire drill during the performance in a Broadway theatre would be ineffective mainly because

(A) the audience is in the theatre for pleasure and would resent interruption of the performance
(B) the members of the audience are transients who do not return to this theatre regularly
(C) the members of the audience have easy access to many exits in the theatre permitting speedy and safe evacuation of the building
(D) theatres are usually of fireproof construction and unlikely to catch fire.

3. "The same care should be taken to avoid damage when extinguishing a fire in the home of a poor family as would be taken in the home of a wealthy family." The best justification for this statement is that the possessions of a poor family

(A) probably are not covered by fire insurance
(B) sometimes include items of great monetary value
(C) may be very precious to the family
(D) may contain valuable or irreplaceable documents such as birth certificates, military discharge certificates, etc.

4. Smoking in bed is considered to be an unsafe practice because the smoker may

(A) become drowsy and carelessly discard smoldering matches or cigarettes
(B) fall asleep and burn his fingers or lips
(C) fall asleep and set fire to his clothing, the bedding, or furniture
(D) fall asleep and choke from the tobacco smoke and fumes.

5. At a fire in a rooming house, the manager tells the officer in command that all the roomers have evacuated the building. The action which would be most proper for the officer to take in the situation is to

(A) order a search of the building for possible victims
(B) question the manager closely to make certain that everybody has in fact been evacuated
(C) question the roomers to obtain confirmation that all occupants have been evacuated
(D) obtain a list of the roomers and then verify that each one of them is outside of the building.

6. An elevator in a large apartment house became stuck between floors and the Fire Department was called to remove the trapped passengers. Soon after arriving, the officer in command informed the passengers that the Fire Department was present and would start rescue operations immediately. The main reason for informing the passengers of their arrival was to reduce the chances that

(A) another agency would receive credit for the rescue
(B) the Fire Department would be criticized for being slow in responding
(C) the passengers would not cooperate with the rescuers
(D) the passengers would become panic-stricken.

7. The election laws of many states require that general elections are to take place on the first Tuesday following the first Monday in November. The statement that is most accurate is that elections in those states

(A) in some years may be held on November 1st and in other years on November 8th
(B) may not be held on either November 1st or November 8th
(C) in some years may be held on November 1st but never on November 8th
(D) in some years may be held on November 8th but never on November 1st.

8. At a smoky fire in a hotel, members of a ladder company are ordered to search all rooms and remove any persons they find remaining in the area threatened by fire or smoke. While performing this duty, a firefighter is prevented from entering a room by a ferocious dog standing in the doorway. In this situation, the best of the following actions for the firefighter to take is to

(A) look into the room without entering and, if no person can be seen inside, assume that the room is empty
(B) obtain assistance from other firefighters, subdue the dog, enter and search the room
(C) attempt to find the dog's owner so that the dog can be removed and the room entered and searched
(D) call for assistance of the A.S.P.C.A. in removing the dog and then enter and search the room.

9. Sparks given off by welding torches are a serious fire hazard. The best of the following methods of dealing with this hazard is to conduct welding operations

(A) only in fireproof buildings protected by sprinkler systems
(B) only out-of-doors on a day with little wind blowing
(C) only on materials certified to be non-combustible by recognized testing laboratories
(D) only after loose combustible materials have been cleared from the area and with a firefighter standing by with a hose line.

10. A firefighter on the way to report for work early one evening noticed a man loitering near a fire alarm box. The man appeared to be touching the alarm box handle but withdrew his hand as the firefighter approached. When asked by the firefighter to explain his actions, the man denied touching the handle. In this situation, the firefighter should

(A) summon a police officer and have the man arrested
(B) ignore the incident since no crime has been committed
(C) keep the man under surveillance until the man returns to his home
(D) locate the patrol officer on the beat and inform him (or her) of the incident.

11. In New York, schools are required to hold at least 12 fire drills each school year, 8 of which must be conducted between September first and December first. Of the following, the best justification for holding most fire drills during the September 1 - December 1 period is that

(A) most fires occur during that period
(B) pupils are trained in evacuation of their schools at the beginning of the school year

(C) the weather is milder during that period
(D) school attendance is higher at the beginning of the school year.

12. Following a severe smoky fire in the lobby of a hotel, a dead body was found in a room on the eighth floor. The coroner who examined the body found that the victim died from the inhalation of toxic gases and smoke. The most probable explanation of this occurrence is that

(A) the coroner made an error in his findings
(B) the victim inhaled the gases and smoke in the lobby, fled to the eighth floor to escape the fire, and collapsed there
(C) the gases and smoke traveled through vertical openings to the upper stories where they were inhaled by the victim
(D) a second fire occurred simultaneously in the upper stories of the hotel.

13. The basic assumption of fire prevention educational programs is that people frequently

(A) must be forced into obeying fire laws
(B) are unaware of the dangers involved in some of their actions
(C) don't care whether or not their actions are dangerous
(D) assume that fire insurance protects them against all fire loss.

14. The statement that is most accurate is that visibility in buildings on fire

(A) is not a serious problem mainly because fires give off sufficient light to enable firefighters to see clearly
(B) is a serious problem mainly because fires often knock out the electrical system of the building
(C) is not a serious problem mainly because most fires occur during daylight hours
(D) is a serious problem mainly because smoke conditions are often encountered.

15. While operating at a five-alarm fire, a firefighter is approached by an obviously intoxicated man who claims to be a former firefighter and who offers to help put out the fire. In this situation, the best of the following courses of action for the firefighter is to

(A) give the man some easy task to perform
(B) refer the man to the officer in command of the fire

(C) decline the offer of help and ask the man to remain outside of the fire lines
(D) ask to see the man's credentials and if he is a former firefighter as he claims put him to work stretching hose lines.

16. A civilian came into a fire house to report that some adolescents had turned on a hydrant in the neighborhood. After getting the location of the hydrant the firefighter on house watch asked the civilian for his name and was told "Smith." The firefighter then thanked the civilian and proceeded to process the report. The firefighter's actions in this situation were

(A) proper
(B) proper except that the firefighter should not have wasted time asking for the civilian's name
(C) proper except that the firefighter should have attempted to find out whether the civilian's name really was "Smith"
(D) proper except that the firefighter should have obtained the civilian's first name and address as well as his family name.

17. Fire apparatus are not permitted to use their sirens when returning to quarters. Of the following, the main justification for this restriction is that

(A) the chances of being involved in traffic accidents are reduced
(B) there is no need for prompt return to quarters after a fire has been extinguished
(C) apparatus return to quarters by less crowded streets than are used when responding to alarms
(D) the officer in command is better able to exchange radio messages with the fire alarm dispatcher.

18. A fire escape platform is attached to a building flush with the window sill leading onto it. The most important reason for this arrangement is that it reduces the chance that

(A) intruders will utilize the fire escapes for illegal purposes since they will be visible from inside the apartment
(B) tenants will place garbage or other obstructions on the platform since these objects will be visible from inside the apartment
(C) persons using the fire escape in an emergency will trip as they leave the apartment
(D) sagging or loosening of the supports will occur without coming to the attention of persons occupying the apartment.

19. A civilian, on his way home from work one evening, hears an alarm ringing and sees water running out of a sprinkler discharge pipe on the side of a building. No smoke or other indications of fire are seen. There is a padlock on the entrance to the building, no fire alarm box is in sight, and a firehouse is located 1½ blocks away. In this situation, it would be most proper for the civilian to

(A) take no action since there is no evidence of fire or other emergency
(B) attempt to ascertain the name of the building's owner and notify him of the situation
(C) send an alarm from the closest fire alarm box
(D) go to the fire house and inform the firefighter at the desk of the situation.

20. Running a hose line between two rungs of a ladder into a building on fire is generally considered to be a

(A) good practice mainly because resting the hose on the rungs of the ladder reduces the load that must be carried by the firefighters
(B) poor practice mainly because the maneuverability of the ladder is reduced
(C) good practice mainly because the weight of the hose line holds the ladder more securely against the side of the building
(D) poor practice mainly because the ladder may be damaged by the additional weight of the hose line filled with water.

21. Suppose that you are the firefighter on housewatch duty when a woman rushes into the fire house and reports that a teen-age gang is assaulting a man on the street several blocks away. In this situation the best of the following courses of action for you to take first is to

(A) notify the Police Department of the reported incident
(B) go to the scene of the disturbance and verify the woman's report
(C) suggest to the woman that she report the matter to the Police Department
(D) notify your fellow firefighters of the incident and all go to the aid of the man being assaulted.

22. A firefighter on the way to work is standing on a subway platform waiting for a train. Suddenly a man standing nearby collapses to the floor, showing no sign of breathing. In this situation, the firefighter should

(A) run to the nearest telephone and summon an ambulance

(B) designate a responsible person to look after the victim and then continue on the way to the firehouse
(C) administer first-aid measures to restore breathing
(D) take no action since the victim is obviously dead.

23. Fire companies often practice methods of firefighting and life saving in public places. The best justification for this practice is that it

(A) provides facilities for practicing essential operations which are not readily available in Fire Department installations
(B) impresses the taxpaying public that they are getting their money's worth
(C) attracts youngsters and interests them in firefighting as a career
(D) makes certain that all company equipment is in the best possible working condition.

24. The fire which would generally present the greatest danger to surrounding buildings from flying sparks or brands is a fire in a/an

(A) warehouse storing paper products
(B) factory manufacturing plastics
(C) outdoor lumberyard
(D) open-type parking garage.

25. A Fire Marshal, questioning firefighters at the scene of a suspicious fire, obtains some conflicting statements about details of the fire situation present upon their arrival. The most likely explanation of this conflict is that

(A) the firefighters have not been properly trained to carry out their part in arson investigations
(B) the Fire Marshal did not give the firefighters time to collect and organize their impressions of the fire scene
(C) witnesses to an event see the situation from their own point of view and seldom agree on all details
(D) details are of little importance if there is general agreement on major matters.

26. A firefighter on duty at a theatre who discovers standees obstructing aisles should immediately

(A) report the situation to his or her superior officer
(B) order the standees to move out of the aisles
(C) ask the theatre manager to correct the situation
(D) issue a summons to the usher assigned to that area.

27. At a fire on the fourth floor of an apartment house, the first engine company to arrive advanced a hose line up the stairway to the third floor before charging the hose with water. The main reason that the firefighters delayed charging their line is that an empty line

(A) is less likely to whip about and injure firefighters
(B) is easier to carry
(C) won't leak water
(D) is less subject to damage.

28. Suppose the owner of a burning tenement building complains that, although the fire is located on the first floor, firefighters are chopping holes in the roof. Of the following, the most appropriate reason you can give for their action is that the fire can be fought most effectively by permitting

(A) smoke and hot gases to escape
(B) firefighters to attack the fire from above
(C) firefighters to gain access to the building through the holes
(D) immediate inspection of the roof for extension of the fire.

29. The Fire Department always endeavors to purchase the best apparatus and equipment and maintain them in the best condition. The main justification for this policy is that

(A) public confidence in the department is increased
(B) failure of equipment at a fire may have serious consequences
(C) replacement of worn out parts is often difficult
(D) the dollar cost to the department is less in the long run.

30. The statement about smoke which is most accurate is that smoke is

(A) irritating but not dangerous in itself
(B) irritating and dangerous only because it may reduce the oxygen content of the air breathed
(C) dangerous because it may reduce the oxygen content of the air breathed and often contains toxic gases
(D) dangerous because it supports combustion.

31. Suppose that you are a firefighter making a routine inspection of a rubber goods factory. During the inspection you discover some minor violation of the Fire Prevention Code. When you call these violations to the attention of the factory owner, he becomes annoyed and tells you that he is the personal friend of high officials in the Fire Depart-

ment and the city government. Under these circumstances, the best course for you to follow is to

(A) summon a police officer to arrest the owner for attempting to intimidate a public official performing his (or her) duty
(B) make a very thorough inspection and serve summonses for every possible violation of the Fire Department Code
(C) ignore the owner's remarks and continue the inspection in your usual manner
(D) try to obtain from the owner the names and positions of his friends.

32. It has been suggested that property owners should be charged a fee each time the Fire Department is called to extinguish a fire on their property. Of the following, the best reason for rejecting this proposal is that

(A) delay in calling the Fire Department may result
(B) many property owners don't occupy the property they own
(C) property owners may resent such a charge as they pay real estate taxes
(D) it may be difficult to determine on whose property a fire started.

33. Standpipe systems of many bridges are the dry pipe type. A dry pipe system has no water in the pipes when not in use; when water is required, it is necessary first to pump water into the system. The main reason for using a dry standpipe system is to prevent

(A) corrosion of the pipes
(B) freezing of water in the pipes
(C) waste of water through leakage
(D) strain on the pumps.

34. Assume that you are a firefighter on your way home after completing your tour of duty. Just as you are about to enter the subway, a man runs up to you and reports a fire in a house located five blocks away. You recognize the man as a simple-minded but harmless person who frequently loiters around fire houses and at fires. Of the following, the best action for you to take is to

(A) run to the house to see if there really is a fire
(B) call in an alarm from a nearby telephone
(C) ignore the report because of the man's mental condition
(D) call the police and ask that a radio patrol car investigate the report.

35. Sometimes a piece of apparatus is ready to leave the fire station before all members are completely dressed and equipped. In order to avoid delay, these firefighters finish dressing on the way to the fire. This practice, although sometimes necessary, is undesirable mainly because

(A) a poor impression is made on the public
(B) the firefighters are not able to size up the situation as they approach the fire
(C) the possibility of dropping equipment from the moving apparatus is increased
(D) the danger of injury to the firefighters is increased.

The following five (5) questions numbered 36 through 40 are based upon the following paragraphs:

1. The sizes of living rooms shall meet the following requirements:

a. In each apartment there shall be at least one living room containing at least 120 square feet of clear floor area, and every other living room except a kitchen shall contain at least 70 square feet of clear floor area.

b. Every living room which contains less than 80 square feet of clear floor area or which is located in the cellar or basement shall be at least 9 feet high and every other living room 8 feet high.

2. Apartments containing three or more rooms may have dining bays, which shall not exceed 55 square feet in floor surface area and shall not be deemed separate rooms or subject to the requirements for separate rooms. Every such dining bay shall be provided with at least one window containing an area at least one-eighthof the floor surface area of such dining bay.

36. The minimum volume of a living room, other than a kitchen, which meets the minimum requirements of the above paragraph is one that measures

(A) 70 cubic feet
(B) 80 cubic feet
(C) 630 cubic feet
(D) 640 cubic feet.

37. A builder proposes to construct an apartment house containing an apartment consisting of a kitchen which measures 10 feet by 6 feet, a room 12 feet by 12 feet and one 11 feet by 7 feet. This apartment

(A) does not comply with the requirements of the paragraph

(B) complies with the requirements of the paragraph provided that it is not located in the cellar or basement
(C) complies with the requirements of the paragraph provided that the height of the smaller rooms is at least 9 feet
(D) may or may not comply with the requirements of the paragraph depending upon the clear floor area of the kitchen.

38. The definition of the term "living room" that is most in accord with its meaning in the above paragraph is

(A) a sitting room or parlor
(B) the largest room in an apartment
(C) a room used for living purposes
(D) any room in an apartment containing 120 square feet of clear floor area.

39. Assume that one room in a four-room apartment measures 20 feet by 10 feet and contains a dining bay 8 feet by 6 feet. According to the above paragraphs, the dining bay must be provided with a window measuring at least

(A) 6 sq. ft.
(B) 7 sq. ft.
(C) 25 sq. ft.
(D) 55 sq. ft.

40. Kitchens, according to the preceding paragraph

(A) are not considered "living rooms"
(B) are considered "living rooms" and must, therefore, meet the height and area requirements of the paragraph
(C) are considered "living rooms" but they need not meet either the height or area requirements of the paragraph
(D) are considered "living rooms" but they need meet only the height requirements, not the area requirements, of the paragraph.

The following five questions numbered 41 through 45 are based upon the following paragraph:

Cotton fabrics treated with the XYZ Process have features which make them far superior to any previously known flame-retardant-treated cotton fabrics. XYZ Process treated fabrics are durable to repeated laundering and dry cleaning; are glow resistant as well as flame resistant; when exposed to flames or intense heat form tough, pliable, and protective chars, are inert physiologically to persons handling or exposed to the fabric; are only slightly heavier than untreated fabrics;

and are susceptible to further wet and dry finishing treatments. In addition, the treated fabrics exhibit little or no adverse change in feel, texture, and appearance, and are shrink-, rot- and mildew-resistant. The treatment reduces strength only slightly. Finished fabrics have "easy care" properties in that they are wrinkle resistant and dry rapidly.

41. It is most accurate to state that the author, in the above paragraph, presents

(A) facts but reaches no conclusion concerning the value of the process
(B) his conclusion concerning the value of the process and facts to support his conclusion
(C) his conclusion concerning the value of the process unsupported by facts
(D) neither facts nor conclusions, but merely describes the process.

42. The article for which the XYZ Process would be most suitable is

(A) nylon stockings
(B) woolen shirt
(C) silk tie
(D) cotton bedsheet.

43. The aspect of the XYZ Process which is <u>not</u> discussed in the above paragraph is its effects on

(A) costs
(B) washability
(C) wearability
(D) the human body.

44. The main reason for treating a fabric with the XYZ Process is to

(A) prepare the fabric for other wet and dry finishing treatment
(B) render it shrink-, rot- and mildew-resistant
(C) increase its weight and strength
(D) reduce the chance that it will catch fire.

45. What would be considered a minor drawback of the XYZ process is that it

(A) forms chars when exposed to flame
(B) makes fabrics mildew-resistant
(C) adds to the weight of fabrics
(D) is compatible with other finishing treatments.

The following four questions, numbered 46 through 49, are based upon the following paragraph:

Language performs an essentially social function; it helps us to get along together, to communicate and achieve a great measure of concerted action. Words are signs which have significance by convention, and those people who do not adopt the conventions simply fail to communicate. They do not "get along" and a social force arises which encourages them to achieve the correct associations. By "correct" is meant as used by other members of the social group. Some of the vital points about language are brought home to an Englishman when visiting America, and vice versa, because our vocabularies are nearly the same - but not quite.

46. "Communicate," as that word is used in the above paragraph, means to

(A) make ourselves understood
(B) send written messages
(C) move other persons to concerted action
(D) use language in its traditional or conventional sense.

47. Usage of a word is "correct," as that term is defined in the above paragraph, when the word is used as it is

(A) defined in standard dictionaries
(B) used by the majority of persons throughout the world who speak the same language
(C) used by the majority of educated persons who speak the same language
(D) used by other persons with whom we are associating.

48. In the above paragraph, the author is concerned primarily with the

(A) meaning of words
(B) pronunciation of words
(C) structure of sentences
(D) origin and development of language.

49. According to the above paragraph, the main language problem of an Englishman, while visiting America, stems from the fact that an Englishman

(A) uses some words that have different meanings for Americans
(B) has different social values than the Americans
(C) has had more exposure to non-English speaking persons than Americans have had
(D) pronounces words differently than Americans do.

The following four questions, numbered 50 through 53, are based upon the following paragraph:

Whenever a social group has become so efficiently organized that it has gained access to an adequate supply of food and has learned to distribute it among its members so well that wealth considerably exceeds immediate demands, it can be depended upon to utilize its surplus energy in an attempt to enlarge the sphere in which it is active. The structure of ant colonies renders them particularly prone to this sort of expansionist policy. With very few exceptions, ants of any given colony are hostile to those of any other community, even of the same species, and this condition is bound to produce preliminary bickering among colonies which are closely associated.

50. According to the above paragraph, a social group is wealthy when it

(A) is efficiently organized
(B) controls large territories
(C) contains energetic members
(D) produces and distributes food reserves.

51. According to the above paragraph, the structure of an ant colony is its

(A) social organization
(B) nest arrangement
(C) territorial extent
(D) food-gathering activities.

52. It follows from the preceding paragraph that the <u>least</u> expansionist society would be one that has

(A) great poverty generally
(B) more than sufficient wealth to meet its immediate demands
(C) great wealth generally
(D) wide inequality between its richest and poorest members.

53. According to the preceding paragraph, an ant is generally hostile <u>except</u> to other

(A) insects
(B) ants
(C) ants of the same species
(D) ants of the same colony.

The following three questions, numbered 54 to 56, relate to the information given in the following paragraph and are to be answered in accordance with this information:

Fire records indicate that about 10 percent of the total value of fire loss in the United States is due to fires caused by heating equipment, and that insufficient clearance to combustible materials is an important factor in a large percentage of these fires. The need for adequate clearance is emphasized at periods of severe cold weather, when heating equipment is run at full capacity, by a marked increase in the number of fires.

54. In a given year in the United States 2,000,000 fires caused 10,000 deaths and destroyed $500,000,000 worth of property. We may conclude from the paragraph that heating equipment caused

(A) 200,000 fires
(B) fires resulting in the death of 1,000 persons
(C) fires resulting in the destruction of $50,000,000 worth of property
(D) 200,000 fires resulting in the death of 1,000 persons and the destruction of $50,000,000 worth of property.

55. By "clearance to combustible materials," as used in the above paragraph, is meant

(A) testing laboratory approval for heating equipment
(B) maintaining heating equipment in clean condition
(C) keeping sufficient space between heating equipment and inflammable material
(D) obtaining appropriate permits before installing the heating equipment.

56. The month-to-month changes in the number of fires caused by heating equipment is most affected by the

(A) season of the year
(B) amount of clearance to combustible materials
(C) condition of the equipment
(D) long term trend in fire statistics.

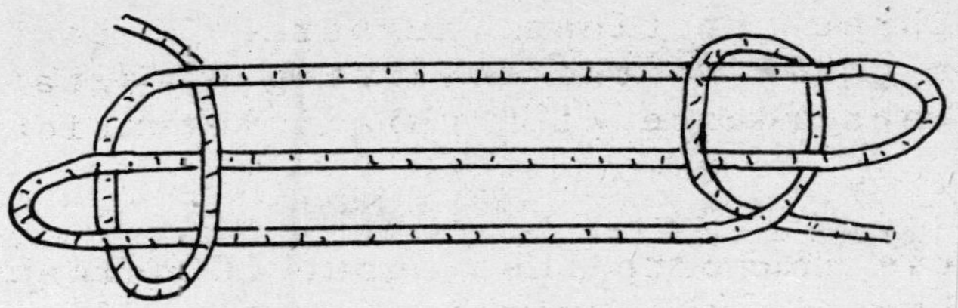

57. The knot shown in the above diagram would best be used to

(A) form a noose for hoisting tools
(B) wrap around fire victims to lower them to the ground
(C) shorten the effective length of a piece of rope
(D) lash two ladders together to extend their lengths.

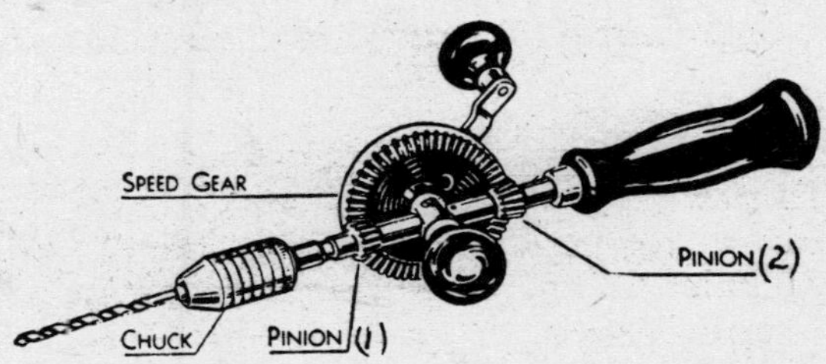

58. The function of pinion gear (2) in the hand drill shown in the above diagram is to

(A) increase the speed of the chuck
(B) keep the speed gear from wobbling
(C) double the turning force on the chuck
(D) allow reverse rotation of the speed gear.

59. Assume that two identical insulated jugs are filled with equal quantities of water from a water tap. A block of ice is placed in one jug and the same quantity of ice, chopped into small cubes, is placed in the other jug. The statement that is most accurate is that the water in the jug containing the chopped ice, compared to the water in the other jug, will be chilled

(A) faster but to a substantially higher temperature
(B) faster and to approximately the same temperature
(C) slower but to a substantially higher temperature
(D) slower and to approximately the same temparature.

Questions numbered 60 through 62 relate to the following diagrams of the dials of a water meter which registers water consumption in cubic feet.

60. The maximum quantity that can be registered directly by the above dials is most nearly

(A) 9,000 cu. ft.
(B) 10,000 cu. ft.
(C) 99,000 cu. ft.
(D) 100,000 cu. ft.

61. The reading of the meter, as shown above, that is most accurate is

(A) 1,465 cu. ft.
(B) 9,047 cu. ft.
(C) 90,465 cu. ft.
(D) 91,475 cu. ft.

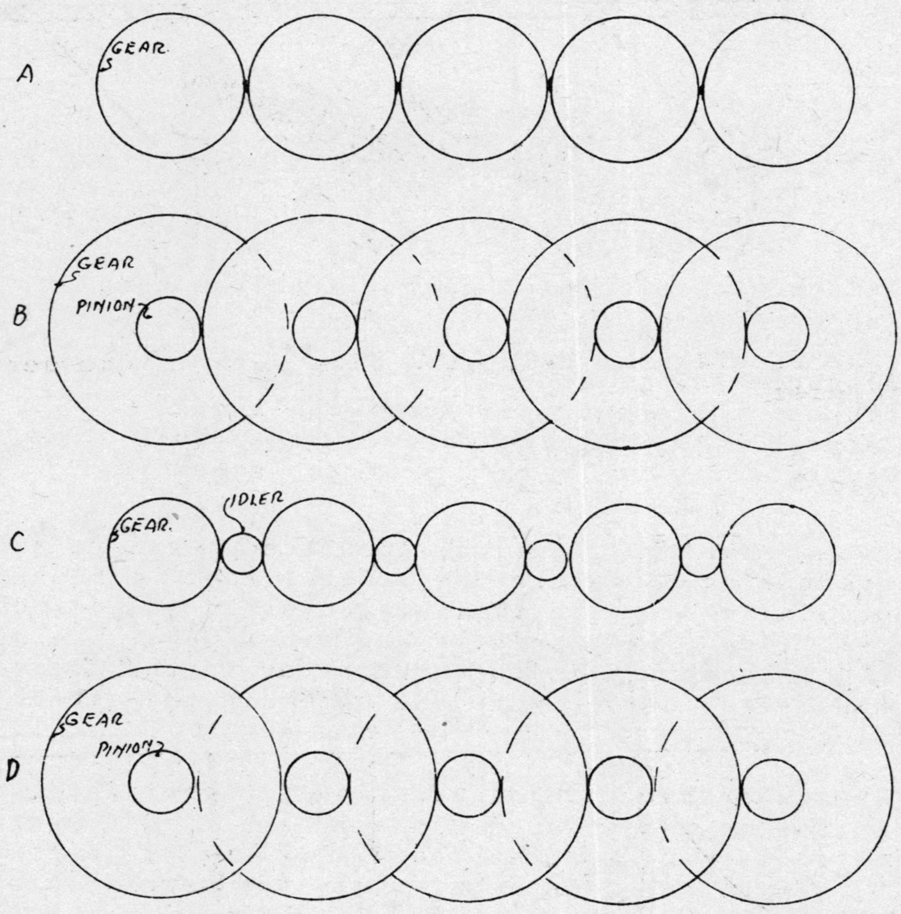

62. The system of gears which could be used to drive the meter shown above is

(A) (B) (C) (D)

63. The long pole and hook shown in the above sketch is called a pike pole. Firefighters sometimes push the point and hook through plaster ceilings and then pull the ceiling down. Of the following, the most likely reason for this practice is to

(A) let heat and smoke escape from the room
(B) trace defective electric wiring through the house
(C) see if hidden fire is burning above the ceiling
(D) remove combustible material which will provide fuel for the fire.

64. Question 64 refers to the figures shown below.

The firefighter using the pike pole in the best way is firefighter

(A) A (C) C
(B) B (D) D.

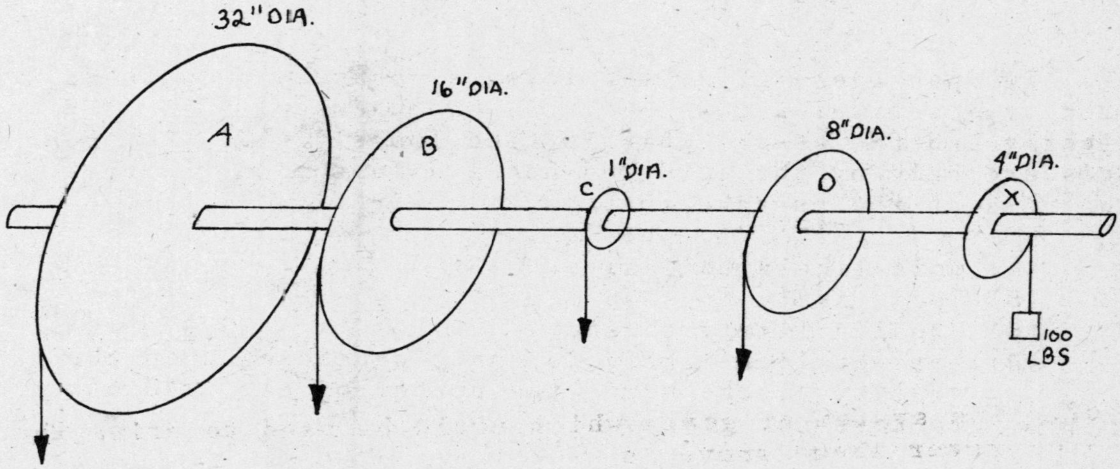

65. In the above diagramed gang of pulleys, all of which are fixed to the common shaft, the 100 lbs. weight attached to

pulley X will be balanced by a weight of 25 lbs. properly attached to

(A) pulley A (C) pulley C
(B) pulley B (D) pulley D.

66. Suppose that the same quantity of water is placed in the cup and bowl pictured above and both are left on a table. The time required for the water to evaporate completely from the cup, compared to the bowl, would be

(A) longer
(B) shorter
(C) equal
(D) longer or shorter depending upon the temperature and humidity in the room.

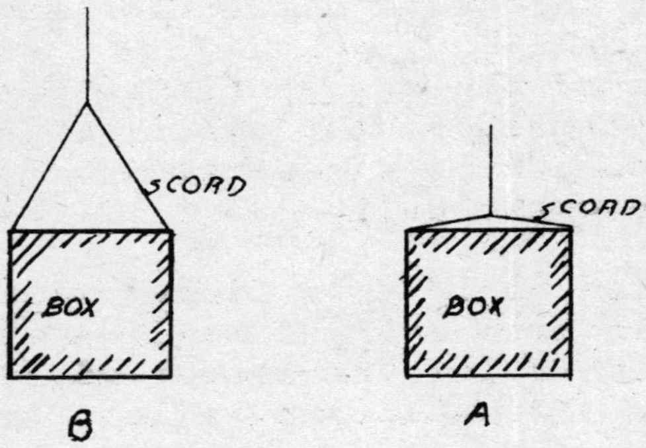

67. Two packages, identical in size, shape and contents are tied by identical pieces of cord. One package, (A), is tied tightly and the other, (B), is tied loosely. If the packages are held by the cord and handled in the same manner, the cord on the package that is held tightly, (A), is

(A) more likely to snap
(B) less likely to snap
(C) equally likely to snap
(D) more or less likely to snap, depending upon the quality of the cord, the number of ply, and other factors not given.

68. Some tools are known as all-purpose tools because they may be used for a great variety of purposes; others are called special purpose tools because they are suitable only for

a particular purpose. Generally, an all-purpose tool, compared to a special tool for the same purpose, is

(A) cheaper
(B) less efficient
(C) safer to use
(D) simpler to operate.

69. During a snowstorm, a passenger car gets stuck in the snow. It is observed that the rear wheels are spinning in the snow and the front wheels are not turning. The statement which best explains why the car is not moving is that

(A) moving parts of the motor are frozen or blocked by the ice and snow
(B) the front wheels are not receiving power because of a defective or malfunctioning transmission
(C) the rear wheels are not obtaining sufficient traction because of the snow
(D) the distribution of the power to the front and rear wheels is not balanced.

70. When the inside face of a rubber suction cup is pressed against a smooth wall it usually remains in place because of the

(A) force of molecular attraction between the rubber and the wall
(B) pressure of the air on the rubber cup
(C) suction caused by the elasticity of the rubber
(D) static electricity generated by friction between rubber and the wall.

71. An observation balloon is filled with just sufficient helium to fully inflate its gas bag. A second identical balloon is filled with twice the amount of helium as the first. The lifting power of the second balloon, compared to the first, is

(A) greater mainly because it contains more helium
(B) greater mainly because it contains helium under higher pressure
(C) equal mainly because the gas bag has the same cubic capacity
(D) less mainly because the additional helium adds to the weight of the balloon without increasing the cubic capacity of the gas bag.

72. When a standing subway passenger pulls the metal "Safety Hand Strap," onto which he is holding, it swings toward him. When he releases it, it swings back to its original position. The most accurate explanation of why the handle returns to its original position is that it is

(A) counterweighted

(B) responding to centrifugal forces acting upon it
(C) spring actuated
(D) well-balanced.

73. The self-contained gas mask consists of a tank containing breathable air which is supplied to the user by means of flexible tubes. This type of gas mask would be <u>least</u> effective in an atmosphere containing

(A) insufficient oxygen to sustain life
(B) gases irritating to the skin
(C) gases which cannot be filtered
(D) a combination of toxic gases.

74. At a fire during very cold weather, a firefighter who was ordered to shut down a hose line left it partly open so that a small amount of water continued to flow out of the nozzle. Leaving the nozzle partially opened in this situation was a

(A) good practice mainly because the hose line can be put back into action more quickly
(B) poor practice mainly because the escaping water will form puddles which will freeze on the street
(C) good practice mainly because the water in the hose line will not freeze
(D) poor practice mainly because unnecessary water damage to property will result.

75. Some sprinkler systems are supplied by water from a single source; others are supplied by water from two different sources. Generally, firefighters consider the two source system preferable to the single source system because the two source system

(A) is cheaper to install
(B) is less likely to be put out of operation
(C) is capable of delivering water at higher pressures
(D) goes into operation faster.

76. Firefighters must often chop holes in roof planking to let smoke escape from burning buildings. When chopping these holes they try to place their axe-cuts just alongside the heavy supporting beams underneath the planking. Of the following, the most likely reason for positioning the cut just to the side of such beams is that

(A) the planking is thinnest at this point
(B) there are no nails in this area to dull the axe
(C) the axe is less likely to "bounce"
(D) the beam provides a strong support for the firefighters using the axe.

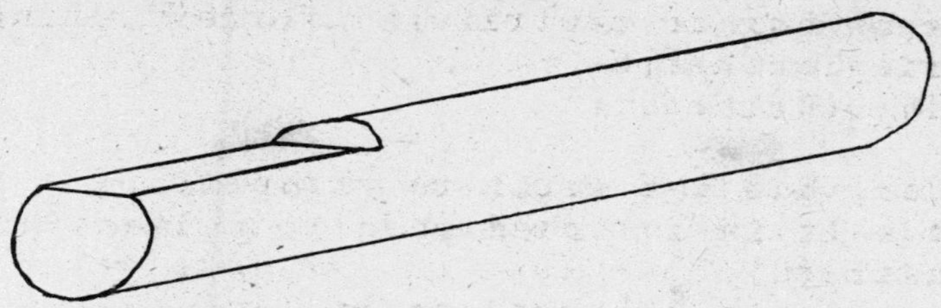

77. The function of the flat surface machined into the shaft in the diagram above is to

(A) prevent slippage of a pulley positioned at this end on the shaft
(B) provide a non-skid surface to hold the shaft steady as it is machined
(C) prevent the shaft from rolling about when it is placed on a flat work bench
(D) reveal sub-surface defects in the shaft.

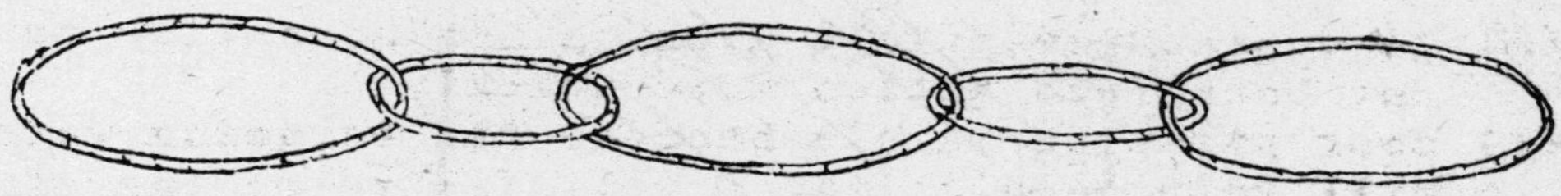

78. A chain consists of links 5" long and ¼" thick alternating with links 2" long and ¼" thick as shown in the above diagram. (Length is measured end-to-end.) If the chain is to be attached to hooks at both ends, the maximum distance that can be spanned by a chain consisting of 20 links is most nearly

(A) a distance which cannot be determined unless it is known whether the chain starts with a large link or a small link
(B) 70 inches
(C) 65 inches
(D) 60 inches.

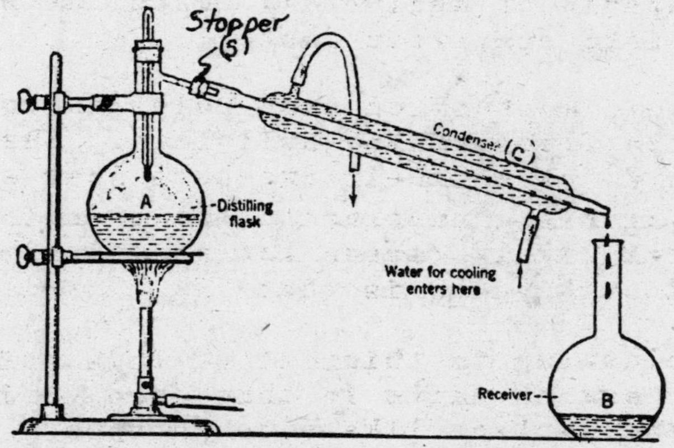

79. The above diagram shows a system for distilling pure

water (flask B) from a salt water solution (flask A). The statement that best explains why there is no salt in the water in flask (B) is that

(A) water, but not salt, is vaporized
(B) the salt is absorbed when it passes through condenser (C)
(C) the salt is filtered at the stopper (S)
(D) only pure water is used in the condenser.

80. The diagram below represents the storage space of a fire engine.

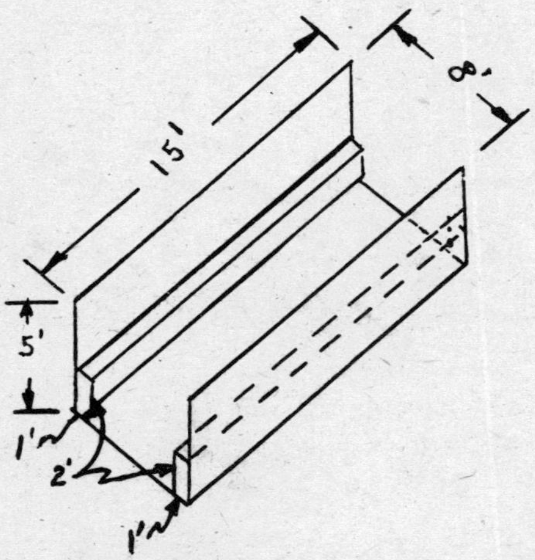

The amount of space available for the storage of hose in the fire engine above is most nearly

(A) 40 cubic feet
(B) 75 cubic feet
(C) 540 cubic feet
(D) 600 cubic feet.

81. If a piece of rope 100 feet long is cut so that one piece is two-thirds as long as the other piece, the length of the longer piece must be

(A) 60 feet
(B) 66-2/3 feet
(C) 70 feet
(D) 75 feet.

82. A water tank has a discharge valve which is capable of emptying the tank, when full, in two hours. It also has an inlet valve which can fill the tank, when empty, in four hours and a second inlet valve which can fill the tank, when empty, in six hours. If the tank is full, and all three valves are opened fully, with water flowing through each valve to capacity, the tank will be emptied in

(A) 2 hours
(B) 6 hours
(C) 12 hours
(D) a period of time which cannot be determined from the information given.

83. In the accompanying sketch of a block and fall, if the end of the rope P is pulled so that it moves one foot, the distance the weight will be raised is

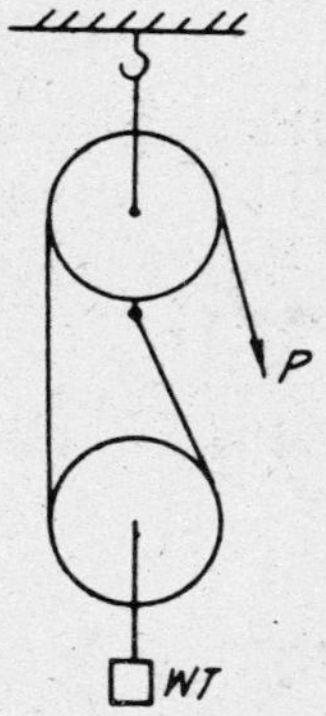

(A) 1/2 ft.
(B) 1 ft.
(C) 1-1/2 ft.
(D) 2 ft.

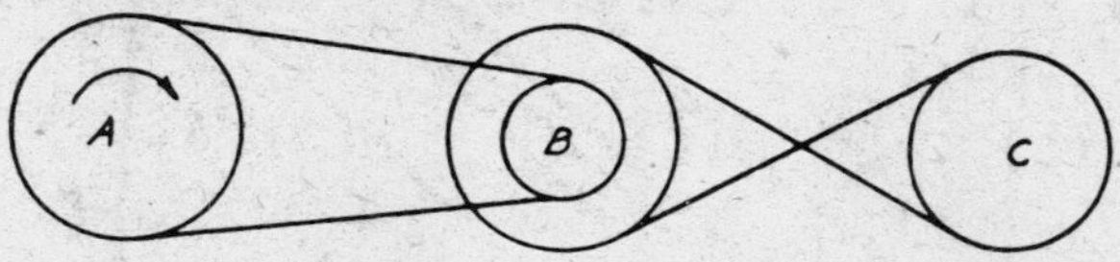

84. The above sketch diagrammatically shows a pulley and belt system. If pulley A is made to rotate in a clockwise direction, then pulley C will rotate

(A) faster than pulley A and in a clockwise direction
(B) slower than pulley A and in a clockwise direction
(C) faster than pulley A and in a counter-clockwise direction
(D) slower than pulley A and in a counter-clockwise direction.

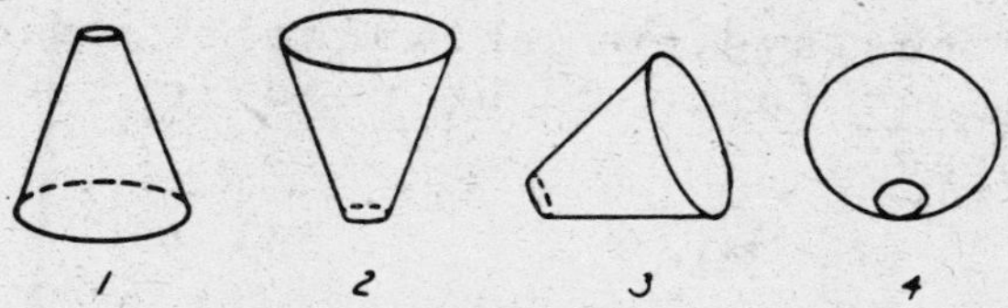

85. The above diagrams show four positions of the same object. The position in which this object is most stable is

(A) 1
(B) 2
(C) 3
(D) 4.

86. The accompanying sketch diagrammatically shows a system of meshing gears with relative diameters as drawn. If

gear 1 is made to rotate in the direction of the arrow, then the gear that will turn fastest is numbered

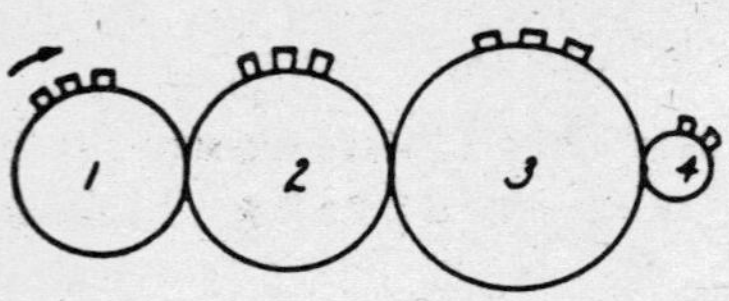

(A) 1
(B) 2
(C) 3
(D) 4.

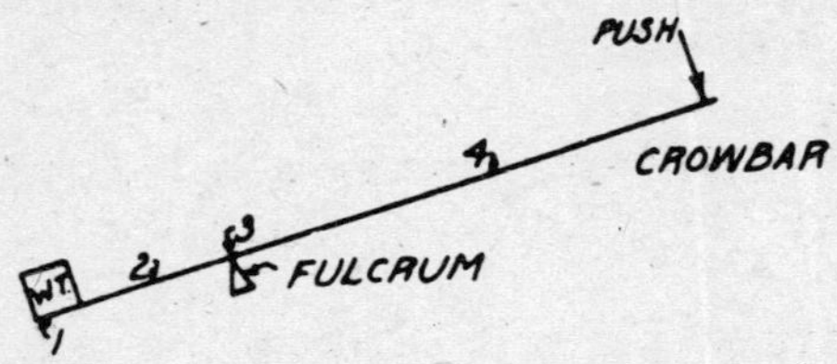

87. The above sketch shows a weight being lifted by means of a crowbar. The point at which the tendency for the bar to break is greatest is

(A) 1 (B) 2 (C) 3 (D) 4.

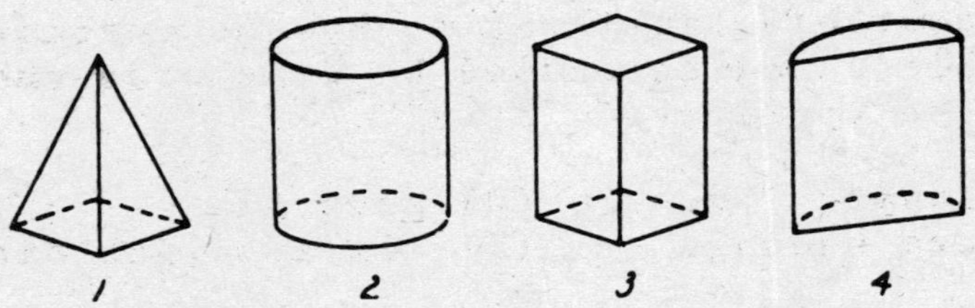

88. The above sketches show four objects which weigh the same but have different shapes. The object which is most difficult to tip over is numbered

(A) 1 (B) 2 (C) 3 (D) 4.

89. An object is to be lifted by means of a system of lines and pulleys. Of the systems shown below, the one which would require the <u>greatest</u> force to be used in lifting the weight is the one numbered

(A) 1 (B) 2 (C) 3 (D) 4.

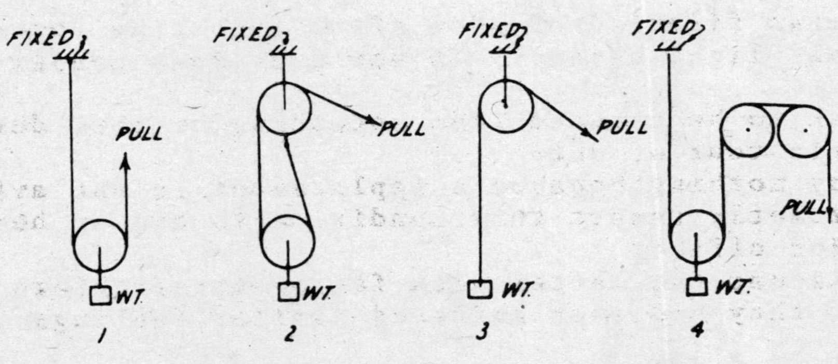

90. During the course of fighting a fire a firefighter encounters an unconscious middle aged man who is bleeding profusely. It would be most proper for the firefighter to administer first aid by initially

(A) attempting to bring him back to consciousness
(B) calling for medical assistance and awaiting its arrival
(C) administering artificial respiration
(D) calling for medical assistance and seeking out the source of the bleeding.

91. The firefighter who encounters a victim of shock should give first aid to the victim by

(A) covering him with a blanket in an attempt to raise his body temperature
(B) administering artificial respiration
(C) loosening his clothing and removing his shoes and stockings
(D) turning him so that he lies on his right side.

92. Modern acceptable first aid technique recommends that wherever possible bleeding caused by punctures should be controlled by the application of a

(A) tourniquet
(B) pressure bandage
(C) coagulant
(D) figure eight bandage.

93. A newly appointed firefighter while on duty becomes involved in an accident. Faced with this problem, it would be most appropriate for the firefighter's superior officer to first ascertain

(A) if the firefighter is accident prone
(B) the cause of the accident
(C) if the orientation program stresses employee safety sufficiently
(D) if the firefighter has ever been involved in an accident previously which was covered by workmen's compensation.

94. A veteran firefighter soon after reporting for a tour of duty feels light headed. It would be most proper to

(A) ask to be excused from ascending heights during this tour of duty
(B) say nothing because a replacement is not available
(C) promptly report this condition to his or her superior officer
(D) discuss the matter with fellow firefighters to see if they had ever suffered similar feelings.

95. A firefighter soon after completing a special course of study on safety at work is not convinced that a method taught during the course is the best way to avoid accidents. Upon returning to duty the firefighter should

(A) perform the task as it was taught
(B) perform the task in the manner he or she is convinced is the best way
(C) attempt to gain the support of fellow firefighters in his or her method
(D) request a transfer to other duty where it will not be necessary to perform this task.

96. A fire company responding to alarms in a minority neighborhood continually encounters abuse from residents who throw bottles, bricks and other refuse at the firefighters while they are fighting fires or answering alarms. The most appropriate course of action for the officer in charge of the fire company to take during one of these incidents is to

(A) assign a number of firefighters to protect the others while they are performing their duties
(B) summon police protection
(C) direct that the firefighters discourage the incidents by having them throw the objects back at the residents
(D) do nothing because this is part of the job of a firefighter.

97. All of the following actions but one would be helpful in gaining support for fire fighting activities from the residents of a neighborhood. The one which would not further this endeavor would be

(A) meetings with neighborhood leaders and the clergy
(B) extension of invitations to the residents to visit the firehouse
(C) publicizing how hard firefighters work
(D) support by the firefighters of a community activity.

98. A firefighter on inspection duty is offered a sum of money by a store owner to overlook a violation of the fire laws. The best action to take would be to

(A) order the store owner to clear the violation without delay and to report the incident to his or her superior officer
(B) say nothing to the store owner but return to the firehouse to seek the advice of fellow firefighters

(C) inform the store owner that firefighters do not accept bribes
(D) refuse the money but offer to assist the store owner to clear the violation during his or her off duty hours.

99. All newly appointed firefighters are put through an extensive training course before they are permitted to participate in the actual activity of fighting fires. The most probable reason for this is that

(A) it would be unwise for inept firefighters to be exposed to the scrutiny of the public
(B) the skills necessary for effective firefighting are unique to the job
(C) the orientation training session is necessary to establish the officer-firefighter relationship
(D) newly appointed firefighters must be convinced of the dangers inherent in firefighting.

100. The statement "Modern technology has greatly influenced firefighting techniques" means most nearly that

(A) newly appointed firefighters are better equipped to fight fires than their predecessors because they are better educated
(B) the need for investigating whether or not changes should be made in procedures becomes important when new firefighting equipment is acquired
(C) established successful firefighting techniques should be retained whenever possible
(D) newly acquired equipment always calls for changes in established procedures.

SAMPLE EXAMINATION II - ANSWERS

1.	C	26.	C	51.	A	76.	C
2.	B	27.	B	52.	A	77.	A
3.	C	28.	A	53.	D	78.	D
4.	C	29.	B	54.	C	79.	A
5.	A	30.	C	55.	C	80.	C
6.	D	31.	C	56.	A	81.	A
7.	D	32.	A	57.	C	82.	C
8.	B	33.	B	58.	B	83.	A
9.	D	34.	B	59.	B	84.	C
10.	D	35.	D	60.	D	85.	A
11.	B	36.	C	61.	C	86.	D
12.	C	37.	C	62.	D	87.	C
13.	B	38.	C	63.	C	88.	A
14.	D	39.	A	64.	D	89.	C
15.	C	40.	D	65.	B	90.	D
16.	D	41.	B	66.	A	91.	D
17.	A	42.	D	67.	A	92.	B
18.	C	43.	A	68.	B	93.	B
19.	D	44.	D	69.	C	94.	C
20.	B	45.	C	70.	B	95.	A
21.	A	46.	A	71.	D	96.	B
22.	C	47.	D	72.	C	97.	C
23.	A	48.	A	73.	B	98.	A
24.	C	49.	A	74.	C	99.	B
25.	C	50.	D	75.	B	100.	B

FIREFIGHTER, F. D.

SAMPLE EXAMINATION III

The time allowed for the entire examination is 3½ hours. In order to create the climate of the test to come, that's precisely the time you should allow yourself. Use a watch and keep a record of your time, especially since you may find it convenient to take the test in several sittings.

1. While performing a routine inspection of a factory building, a firefighter is asked a question by the plant manager about a matter which is under the control of the Health Department and about which the firefighter has little knowledge. In this situation, the best course of action for the firefighter to take is to

(A) answer the question to the best of his (or her) ability
(B) tell the manager that he (or she) is not permitted to answer the question because it does not relate to a Fire Department matter
(C) tell the manager that it will be referred to the Health Department
(D) suggest to the manager that he communicate with the Health Department about the matter.

2. A firefighter on duty who answers a departmental telephone should give his or her name and rank

(A) at the start of the conversation, as a matter of routine
(B) only if asked for this information by the caller
(C) only if the caller is a superior officer
(D) only if the telephone message requires the firefighter to take some action.

3. At a recent five alarm fire in the Bronx, several companies from Manhattan were temporarily assigned to occupy the quarters and take over the duties of companies engaged in fighting the fire. The main reason for relocating the Manhattan companies was to

(A) protect the fire houses from robbery or vandalism which might occur if they were left vacant for a long period
(B) provide for speedy response to the fire if additional companies are required

(C) give the Manhattan companies an opportunity to become familiar with the problems of the Bronx area
(D) provide protection to the Bronx area in the event other fires should occur.

4. Two firefighters, on their way to report for duty early one morning, observe a fire in a building containing a supermarket on the street level and apartments on the upper stories. One firefighter runs into the building to spread the alarm to the tenants. The other firefighter runs to a street alarm box two blocks away and sends an alarm. The latter firefighter should then

(A) return to the building which is on fire and help evacuate the tenants
(B) remain at the fire alarm box in order to direct the first fire company that arrives to the location of the fire
(C) look for a telephone in order to call his (or her) own fire company and explain that he and his companion will be late in reporting for duty
(D) look for a telephone in order to call the Health Department and request that an inspector be sent to the supermarket to examine the food involved in the fire.

5. A man found an official NYFD badge and gave it to his young son to use as a toy. The man's action was improper mainly because

(A) it is disrespectful to the Fire Department to use the badge in this manner
(B) the boy may injure himself playing with the badge
(C) an effort should have been made first to locate the owner of the badge before giving it to the boy
(D) the badge should have been returned to the Fire Department.

6. In case of a fire in a U.S. Mail Box the Fire Department recommends that an extinguishing agent which smothers the fire, such as carbon tetrachloride, should be used. Of the following, the most likely reason for not recommending the use of water is that

(A) water is not effective on fires in small tightly enclosed spaces
(B) someone might have mailed chemicals that could explode in contact with water
(C) water may damage the mail untouched by fire so that it could not be delivered
(D) the smothering agent can be put on the fire faster than water can be.

7. Of the following, the main difficulty in the way of obtaining accurate information about the causes of fires is that

(A) firefighters are too busy putting out fires to have time for investigation of the causes of fires
(B) most people have little knowledge about fire hazards
(C) fires destroy much of the evidence which would indicate the causes of the fires
(D) fire departments are more interested in fire prevention than in investigating fires that have already occurred.

8. In an effort to discourage the sending of false alarms and to help apprehend those guilty of this practice it is suggested that the handles of fire alarm boxes be covered with a dye which would stain the hand of a person sending an alarm, and which would not wash off for 24 hours. The dye would be visible only under an ultra-violet light. Of the following, the chief objection to such a device is that it would

(A) require funds that can be better used for other purposes
(B) have no effect on false alarms transmitted by telephone
(C) discourage some persons from sending alarms for real fires
(D) punish the innocent as well as the guilty.

9. Automatic fire extinguishing sprinkler systems are sometimes not effective on fires accompanied by explosions, chiefly because

(A) these fires do not generate enough heat to start sprinkler operation
(B) the pipes supplying the sprinklers are usually damaged by the explosion
(C) fires in explosive materials usually cannot be extinguished by water
(D) sprinkler heads are usually clogged by dust created by the explosion.

10. When a fire occurs in the vicinity of a subway system there is the possibility that water from the firefighters' hose streams will flood underground portions of the subway lines through sidewalk gratings. Of the following methods of reducing this danger, the one that would generally be most suitable is for the officer in command to order his subordinates to

(A) use fewer hose lines and smaller quantities of water than they would ordinarily
(B) attack the fire from positions which are distant from the sidewalk gratings
(C) cover the sidewalk gratings with canvas tarpaulins
(D) advise the subway dispatcher to re-route the subway trains.

11. When responding to alarms, Fire Department apparatus generally follow routes established in advance. The one which would be the least valid justification for this practice is that

(A) motorists living in the area become familiar with these routes and tend to avoid them
(B) the likelihood of collision between two pieces of Fire Department apparatus is reduced
(C) the fastest response is generally obtained
(D) road construction, road blocks, detours, and similar conditions can be avoided.

12. An off-duty firefighter sees, from a distance, a group of teen-age boys set fire to a newspaper and toss the flaming pages into an open window of a building which is being torn down. In this situation, the first action which should be taken by the firefighter is to

(A) send a fire alarm from the closest street alarm box
(B) chase the boys and attempt to catch one of them
(C) investigate whether a fire has been started
(D) call the police from the closest police alarm box or telephone.

13. When responding to an alarm, officers are not to talk to chauffeurs driving the apparatus except to give orders or directions. Of the following, the best justification for this rule is that it

(A) gives the officer an opportunity to make preliminary plans for handling the fire problem
(B) enables the chauffeur to concentrate on driving the apparatus
(C) maintains the proper relationship between the ranks while on duty
(D) permits the officer to observe the chauffeur's skill, or lack of skill, in driving the apparatus.

14. The approved method of reporting a fire by telephone in New York City is first to dial the

(A) central headquarters of the Fire Department
(B) borough headquarters of the Fire Department

(C) local fire station house
(D) telephone operator.

15. Doors in theaters and other places of public assembly usually open outwardly. The main reason for this requirement is, in the event of fire, to

(A) provide the widest possible passageway for escape of the audience
(B) prevent panic stricken audiences from jamming the doors in a closed position
(C) indicate to the audience the safe direction of travel
(D) prevent unauthorized persons from entering the building.

16. Fire prevention inspections should be conducted at irregular hours or intervals. The best justification for this "irregularity" is that it permits the firefighters to

(A) make inspections when they have free time
(B) see the inspected establishments in their normal condition and not in their "dressed-up" condition
(C) avoid making inspections at times which would be inconvenient for the inspected establishments
(D) concentrate their inspectional activities on those establishments which present the greatest fire hazard.

17. Some gas masks provide protection to the user by filtering out from the air certain harmful gases present in the atmosphere. A mask of this type would <u>not</u> be suitable in an atmosphere containing

(A) heavy, black smoke
(B) a filterable gas under pressure
(C) insufficient oxygen to sustain life
(D) more than one filterable harmful gas.

18. Firefighters are instructed never to turn on the gas supply to a house, which was turned off by them during a fire. Of the following, the most important reason for this prohibition is that

(A) the fire may have made the gas meters inaccurate
(B) unburned gas may escape from open gas jets
(C) the utility company's employees may object to firefighters performing their work
(D) firefighters should not do anything which is not directly related to extinguishing fires.

19. A firefighter in uniform, performing inspectional duty, comes upon a group of young men assaulting a police officer.

The firefighter goes to the aid of the police officer and, in the course of the struggle, receives some minor injuries. The action of the firefighter in this situation was

(A) proper, chiefly because members of the uniform forces must "stick together"
(B) improper, chiefly because people in the neighborhood, as a result, might refuse to cooperate with the Fire Department's various programs
(C) proper, chiefly because all citizens have an obligation to assist police officers in the performance of their duty
(D) improper, chiefly because the Fire Department lost the services of the firefighter while the firefighter was recovering from his or her injuries.

20. Members of the Fire Department may not make a speech on Fire Department matters without the approval of the Fire Commissioner. Requests for permission must be accompanied by a copy or summary of the speech. The main reason for this requirement is to

(A) determine whether the member is engaged in political activities which are forbidden
(B) reduce the chance that the public will be misinformed about Fire Department policies or procedures
(C) provide the Department with a list of members who can serve in the Department's speakers' bureau
(D) provide the Department with information about the off-duty activities of members.

21. For a firefighter to straddle a hose line while holding the nozzle and directing water on fires is a

(A) good practice mainly because better balance is obtained by the firefighter
(B) poor practice mainly because the firefighter directing the hose may trip over the hose
(C) good practice mainly because better control over the hose line is obtained
(D) poor practice mainly because the hose might whip about and injure the firefighter.

22. Firefighters are required to wear steel reinforced inner-soles inside their rubber boots. The main purpose of these inner-soles is to

(A) make the boots more durable and long lasting
(B) protect the firefighter's feet from burns from smoldering objects or embers
(C) protect the firefighter's feet from injury from falling objects

(D) protect the firefighter's feet from nails or other sharp objects.

23. Promoting good relations with the public is an important duty of every member of the Fire Department. Of the following, the best way for a firefighter to promote good public relations is to

(A) become active in civic and charitable organizations
(B) be well dressed, clean, and neat on all occasions
(C) write letters to newspapers explaining the reasons for departmental procedures
(D) perform his duties with efficiency, consideration, and courtesy.

24. Firefighters are advised to avoid wearing rings on their fingers. The main reason for this advice is that the rings have a tendency to

(A) be damaged during operations at fires
(B) scratch persons receiving first aid treatment
(C) catch on objects and injure the wearer
(D) scratch furniture and/or other valuable objects.

25. Suppose that you are a firefighter on housewatch duty when a civilian enters the fire house. He introduces himself as a British firefighter visiting the country to study American firefighting methods. He asks you for permission to ride on the fire apparatus when it responds to alarms in order to observe operations at first hand. You know it is against departmental policy to permit civilians to ride apparatus without written permission from headquarters. In this situation you should

(A) refuse the request but suggest that he follow the apparatus in his own car when it responds to an alarm
(B) call headquarters and request permission to permit the visitor to ride the apparatus
(C) refuse the request and suggest that he apply to headquarters for permission
(D) refuse the request and suggest that he should return the next time that the Fire Department holds open house.

26. When operating at a pier fire, firefighters usually avoid driving their apparatus onto the pier itself. The main reason for this precaution is to reduce the possibility that the apparatus will be

(A) delayed in returning to quarters
(B) driven off the end of the pier

(C) destroyed by a fire that spreads rapidly
(D) in the way of the firefighters.

27. Pumpers purchased by the Fire Department are equipped with enclosed cabs. In the past, Fire Department apparatus was the open type, with no cab or roof. The main advantage of the enclosed cab is that it provides

(A) additional storage space for equipment
(B) a place of shelter for firefighters operating in an area of radioactivity
(C) protection for firefighters from weather conditions and injury
(D) emergency first aid and ambulance facilities.

28. Heavy blizzards greatly increase the problems and work of the Fire Department. When such a situation occurs, the Fire Commissioner could reasonably be expected to

(A) order members of the Fire Department to perform extra duty
(B) limit parking on city streets
(C) station firefighters at fire alarm boxes to prevent the sending of false alarms
(D) prohibit the use of kerosene heaters.

29. Regulations of the Fire Department require that when placing hose on a fire wagon care should be taken to avoid bending the hose at places where it had been bent previously. The most important reason for this requirement is that repeated bending of the hose at the same places will cause

(A) kinks in the hose at those places
(B) weakening of the hose at those places
(C) discoloration of the hose at those places
(D) dirt to accumulate and clog the hose at those places.

30. While fighting a fire in an apartment when the occupants are not at home, a firefighter finds a sum of money in a closet. Under these circumstances, the firefighter should turn over the money to

(A) a responsible neighbor
(B) the desk sergeant of the nearest police station
(C) the superintendent of the apartment house
(D) his or her superior officer.

31. "When necessary to remove a cornice, every effort should be made to pull it back on the roof." The most important reason for the direction in the above quotation is that pulling the cornice on the roof rather than dropping it to the street below

(A) requires less time
(B) is safer for the people on the street
(C) makes it possible to re-use the cornice
(D) is less dangerous to firefighters working on the roof.

32. After a fire has been extinguished, a firefighter often remains at the scene after the others have left. Of the following, the main reason for this practice is that this firefighter can

(A) prevent looters from stealing valuables
(B) watch for any rekindling of the fire
(C) search the area for lost valuables
(D) examine the premises for evidence of arson.

33. Firefighters usually attempt to get as close as possible to the seat of a fire so that they can direct their hose streams with accuracy. However, intense heat sometimes keeps them at a distance. What is <u>not</u> a satisfactory method of overcoming this problem is to

(A) have firefighters use some solid object, such as a wall, as a shield
(B) keep firefighters cool by wetting them with small streams of water
(C) use large high pressure streams and operate at a great distance from the fire
(D) use a water spray to break down the heat waves coming from the fire.

34. Suppose that you are an off-duty firefighter driving your car in the mid-town area. As you cross an intersection you hear sirens and, looking back, see fire apparatus approaching. In this situation, the best action for you to take is to

(A) attempt to clear a path for the fire apparatus by driving rapidly and sounding your horn
(B) drive to the next intersection and direct traffic until the apparatus has passed
(C) permit the apparatus to pass, then follow it closely to the fire, sounding your horn as you drive
(D) pull to the curb, permit the apparatus to pass, then continue on your way.

35. Whenever a public performance is given in a theater involving the use of scenery or machinery, a firefighter is assigned to be present. The main reason for this assignment is that

(A) theaters are located in high property value districts
(B) the use of scenery and machinery increases the fire hazard

(C) theatrical districts have heavy traffic, making for slow response of apparatus
(D) emergency exits may be blocked by the scenery or machinery.

Questions numbered 36 through 39 are based upon the following paragraph:

A plastic does not consist of a single substance, but is a blended combination of several. In addition to the resin, it may contain various fillers, plasticizers, lubricants, and coloring material. Depending upon the type and quantity of substances added to the binder, the properties, including combustibility, may be altered considerably. The flammability of plastics depends upon the composition and, as with other materials, upon their physical size and condition. Thin sections, sharp edges, or powdered plastics will ignite and burn more readily than the same amount of identical material in heavy sections with smooth surfaces.

36. According to the above paragraph, all plastics contain a

(A) resin
(B) resin and a filler
(C) resin, filler, and plasticizer
(D) resin, filler, plasticizer, lubricant, and coloring material.

37. The conclusion that is best supported by the above paragraph is that the flammability of plastics

(A) is generally high
(B) is generally moderate
(C) is generally low
(D) varies considerably.

38. According to the above paragraph, "plastics" can best be described as

(A) a trade name
(B) the name of a specific product
(C) the name of a group of products which have some similar and some dissimilar properties
(D) the name of any substance which can be shaped or molded during the production process.

39. In a manufacturing process, large sheets of a particular plastic are cut, buffed, and formed into small tools. The statement that is most in accord with the information in the above paragraph is that

(A) the dust particles of the plastic are more flammable than the tools or the sheets
(B) the plastic tools are more flammable than the dust particles or the sheets
(C) the sheets of the plastic are more flammable than the dust particles or the tools
(D) there is insufficient information to determine the relative flammability of sheets, tools, and dust.

Questions numbered 40 through 42 are based upon the following paragraph:

To guard against overheating of electrical conductors in buildings, an overcurrent protective device is provided for each circuit. This device is designed to open the circuit and cut off the flow of current whenever the current exceeds a predetermined limit. The fuse, which is the most common form of overcurrent protection, consists of a fusible metal element which when heated by the current to a certain temperature, melts and opens the circuit.

40. According to the above paragraph, a circuit which is <u>not</u> carrying an electric current is

(A) an open circuit
(B) a closed circuit
(C) a circuit protected by a fuse
(D) a circuit protected by an overcurrent protective device other than a fuse.

41. As used in the above paragraph, the best example of a conductor is

(A) a metal table which comes in contact with a source of electricity
(B) a storage battery generating electricity
(C) an electrical wire carrying an electrical current
(D) a dynamo converting mechanical energy into electrical energy.

42. According to the above paragraph, the maximum number of circuits that can be handled by a fuse box containing 6 fuses

(A) is 3
(B) is 6
(C) is 12
(D) cannot be determined from the information given in the above paragraph.

Questions numbered 43 through 45 are based upon the following paragraph:

Unlined linen hose is essentially a fabric tube made of closely woven linen yarn. Due to the natural characteristics of linen, very shortly after water is introduced, the threads swell after being wet, closing the minute spaces between them making the tube practically water tight. This type of hose tends to deteriorate rapidly if not thoroughly dried after use or if installed where it will be exposed to dampness or the weather. It is not ordinarily built to withstand frequent service or for use where the fabric will be subjected to chafing from rough or sharp surfaces.

43. Seepage of water through an unlined linen hose is observed when the water is first turned on. From the above paragraph, we may conclude that the seepage

(A) indicates that the hose is defective
(B) does not indicate that the hose is defective provided that the seepage is proportionate to the water pressure
(C) does not indicate that the hose is defective provided that the seepage is greatly reduced when the hose becomes thoroughly wet
(D) does not indicate that the hose is defective provided that the seepage takes place only at the surface of the hose.

44. Unlined linen hose is most suitable for use

(A) as a garden hose
(B) on fire department apparatus
(C) as emergency fire equipment in buildings
(D) in fire department training schools.

45. The use of unlined linen hose would be <u>least</u> appropriate in

(A) an outdoor lumber yard
(B) a non-fireproof office building
(C) a department store
(D) a cosmetic manufacturing plant.

The following four questions, numbered 46 to 49, are based on the information given in the paragraph below and are to be answered in accordance with this information.

The principal value of inspection work is in the knowledge obtained relating to the various structural features of the building and the protective features provided. Knowledge of the location of stairways and elevators, the obstruction provided by merchandise, the danger from absorption of water by baled stock, the potential hazard of rupture of containers such as drums or cylinders, and the location of protective equipment, are all essential features to be noted and later discussed in company school and officer's college.

46. According to the above paragraph, the chief value of inspection work is to gather information which will aid in

(A) fixing responsibility for fires
(B) planning firefighting operations
(C) training new firefighters
(D) obtaining compliances with the Building Code.

47. The object which would be the most help in accomplishing the objective of the inspection as stated in the above paragraph is a

(A) copy of the Building Code
(B) chemical analysis kit
(C) plan of the building
(D) list of the building's tenants.

48. An example of a "structural feature" contained in the above paragraph is the

(A) location of stairways and elevators
(B) obstruction provided by merchandise
(C) danger of absorption of water by baled stock
(D) hazard of rupture of containers such as drums or cylinders.

49. Of the following, the best example of what is meant by a "protective feature," as used in the above paragraph, is

(A) a fire extinguisher
(B) a burglar alarm
(C) fire insurance
(D) a medical first-aid kit.

50. Although the periodic inspection of buildings is required by law, a Fire Department can constructively utilize information gained during these inspections to assist in the

(A) training of its <u>new</u> firefighters
(B) substantiation of budget requests
(C) planning of firefighting operations
(D) promotion of its community activities.

51. A firefighter on inspection duty should behave in a manner that is likely to gain the support of the public that his or her agency serves. The firefighter can accomplish this by engaging in all but one of the following practices. The practice that will not accomplish this is

(A) being courteous and neat in appearance
(B) being prepared to give explanations of the Fire Prevention Code
(C) overlooking minor infractions of the Fire Prevention Code because correcting them would be expensive
(D) performing inspectional duties in a businesslike manner.

52. A firefighter receives a call during the night while off duty from a neighbor who says his house is on fire. The most appropriate action for the firefighter to take <u>first</u> is to

(A) call for the assistance of the Fire Department if it has not been already called
(B) go to the neighbor's home to evaluate the extent of the fire
(C) call other off duty firefighters who live close by to help fight the fire
(D) advise the neighbor that you will be over as soon as you are dressed.

The following four questions, numbered 53 to 56 , relate to the information given in the paragraph below and are to be answered in accordance with this information.

Division commanders shall arrange and maintain a plan for the use of hose wagons to transport members in emergencies. Upon receipt of a call for members, the deputy chief of the division from whom the firefighters are called shall have the designated hose wagon placed out of service and prepared for the transportation of members. Hose wagons shall be placed at central assembly points, and members detailed instructed to report promptly to such locations equipped for fire duty. Hose wagons designated shall remain at regular assignments when not engaged in the transportation of members.

53. Preparation of the hose wagon for this special assignment of transporting of members would most likely involve

(A) checking the gas and oil, air in tires, and mechanical operation of the apparatus
(B) removal of hose lines to make room for the members being transported
(C) gathering of equipment which will be needed by the members being transported
(D) instructing the driver on the best route to be used.

54. Hose wagons used for emergency transportation of members are placed out of service because they

(A) are not available to respond to alarms in their own district
(B) are more subject to mechanical breakdown while on emergency duty
(C) are engaged in operations which are not the primary responsibility of their division
(D) are considered reserve equipment.

55. Of the following, the best example of the type of emergency referred to in the above paragraph is

(A) a firefighter injured at a fire and requiring transportation
(B) a subway strike which prevents firefighters from reporting for duty
(C) an unusually large number of false alarms occurring at one time
(D) a need for additional manpower at a fire.

56. A "central assembly point," as used in the above paragraph, would most likely be a place

(A) close to the place of the emergency
(B) in the geographical center of the division
(C) easily reached by the members assigned
(D) readily accessible to the intersection of major highways.

57. Modern fire houses have automatic hose dryers for drying rubber-lined hose after use. The controls are usually set to provide heat no higher than 25°F. above room temperature. Of the following, the most important reason for not using more heat is that the

(A) wear and tear on the dryers would become excessive
(B) hose dries more thoroughly if it dries slowly
(C) hose would be too hot to handle
(D) rubber lining would deteriorate too rapidly.

58. The best explanation why smoke usually rises from a fire is that

(A) cooler, heavier air displaces lighter, warm air
(B) heat energy of the fire propels smoke upward
(C) suction from the upper air pulls the smoke upward
(D) burning matter is chemically changed into heat energy.

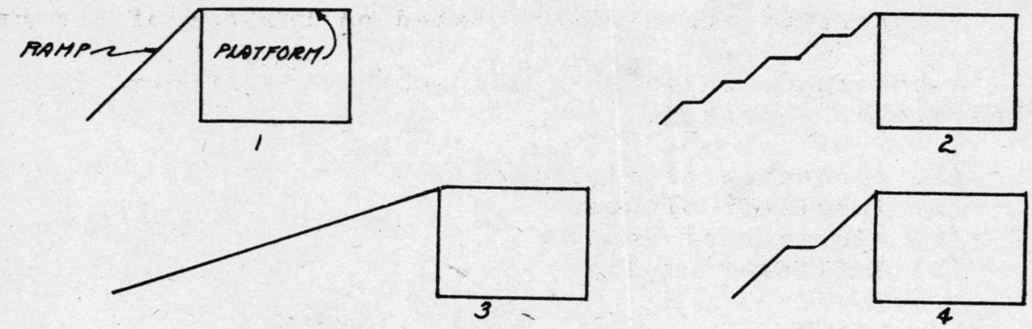

59. The above diagram shows various types of ramps leading to a loading platform. The ramp which would permit the load to be moved up to the platform with the least amount of force is

(A) 1 (B) 2 (C) 3 (D) 4.

60. The practice of racing a car engine to warm it up in cold weather is generally

(A) good mainly because repeated stalling of the engine and drain on the battery is avoided
(B) bad mainly because too much gas is used to get the engine heated
(C) good mainly because the engine becomes operational in the shortest period of time
(D) bad mainly because proper lubrication is not established rapidly enough.

61. Ice on sidewalks can often be melted by sprinkling salt on it. The melting of the ice results from

(A) a chemical reaction between the salt and ice which produces heat
(B) attraction of sun rays by the salt to the ice
(C) lowering of the freezing point of water by the salt
(D) heat of friction caused by persons walking on the salt.

62. Only one of the following four statements relating to the temperature at which water will boil is correct. The correct statement is that

(A) water always boils at the same temperature regardless of pressure
(B) water heated slowly by a low flame will boil at a higher temperature than water heated quickly by a high flame
(C) a large quantity of water will boil at a higher temperature than a small quantity
(D) water heated at sea level will boil at a higher temperature than water heated on the top of a mountain.

63. A substance which is a good conductor of heat is most likely to be a poor

(A) conductor of electricity
(B) insulator of heat
(C) vibrator of sound
(D) reflector of light.

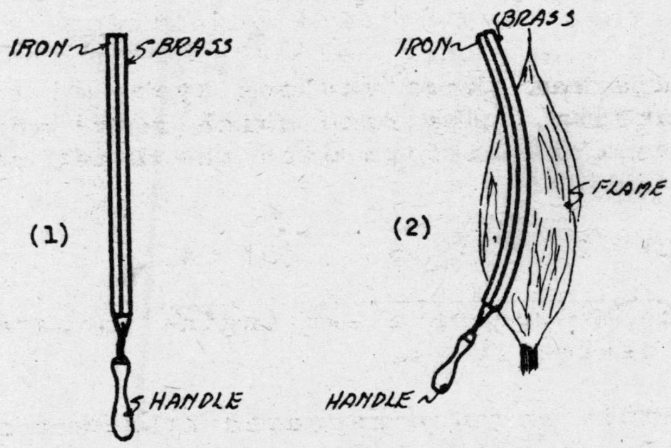

64. The above diagrams show what happens to a bar consisting of iron on one side fastened to brass on the other when it is heated in the flame of a burner. The best explanation of the curvature of the bar in diagram 2 is that

(A) iron expands more than brass when heated
(B) brass expands more than iron when heated
(C) the iron side of the bar was in the hottest part of the flame
(D) the brass side of the bar was in the hottest part of the flame.

65. A load is to be supported from a steel beam by a chain consisting of 20 links and a hook. If each link of the chain weighs 1 pound and can support a weight of 1,000 pounds, and the hook weighs 5 pounds and can support a weight of 5,000 pounds, the maximum load that can be supported from the hook is most nearly

(A) 25,000 lbs.
(B) 5,000 lbs.
(C) 1,000 lbs.
(D) 975 lbs.

66. Ice formation in water pipes often causes the bursting of the pipes because

(A) the additional weight of ice overloads the pipes
(B) water cannot pass the ice block and builds up great pressure on the pipes
(C) the cold causes contraction of the pipes and causes them to pull apart
(D) water expands upon freezing and builds up great pressure on the pipes.

67. Oceanshore areas tend to have less temperature variation between winter and summer extremes than inland areas. Of the following, the best explanation for this observation is that generally

(A) prevailing winds are from the ocean to land in the summer and from land to the ocean in the winter
(B) inland areas have natural vegetation which absorbs summer heat and then releases it in the winter
(C) water adjacent to shoreline areas absorbs summer heat and then releases it in the winter
(D) warm water ocean currents moderate the temperature of land adjacent to the shoreline.

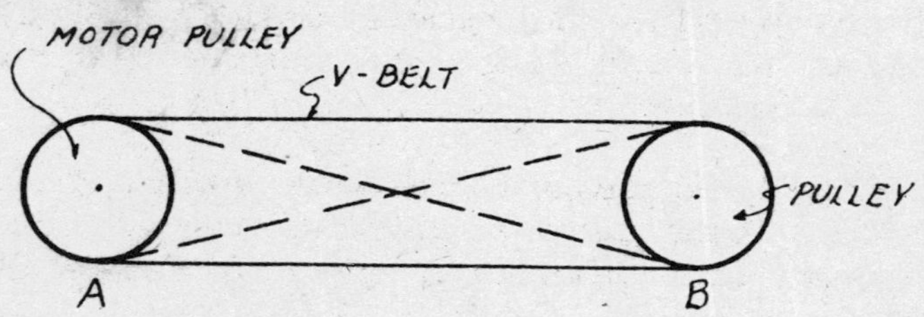

68. In the above diagram, crossing the V belt as shown by dotted lines will result in

(A) pulley A reversing direction
(B) no change in the direction of either pulley
(C) pulley B reversing direction
(D) stoppage of motor.

Questions numbered 69 through 71 are based upon the following diagram:

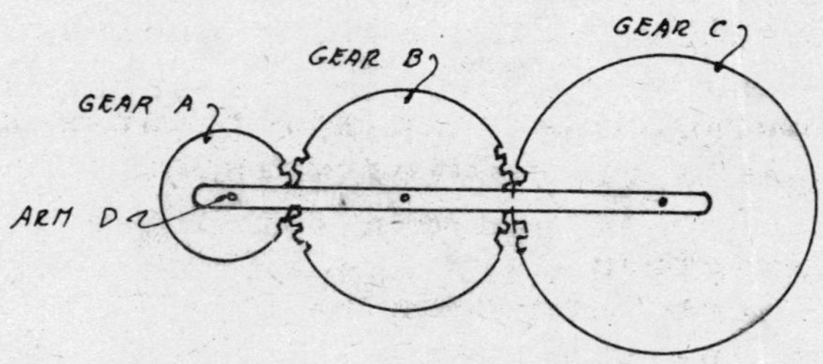

(Assume that the teeth of the gears are continuous all the way around each gear.)

69. Fastening gear A to arm D at another point in addition to its shaft will result in

(A) gear B rotating on its shaft in a direction opposite to gear A
(B) gear C rotating on its shaft in a direction opposite to gear A
(C) arm D rotating around the shaft of gear C
(D) locking of all gears.

70. If gear C is fastened to a supporting frame, not shown, so that it cannot rotate and gear A turns clockwise on its shaft then

(A) gear B will turn counter-clockwise and arm D will turn clockwise around the shaft of gear C
(B) gear B will turn counter-clockwise and arm D will turn counter-clockwise around the shaft of gear C
(C) gear B will turn clockwise and arm D will turn clockwise around the shaft of gear C
(D) gear B will turn clockwise and arm D will turn counter-clockwise around the shaft of gear C.

71. If gear B is fastened to a supporting frame, not shown, so that it cannot rotate and arm D rotates clockwise around the shaft of gear B, then for each complete revolution arm D makes, gear A will make

(A) more than one turn clockwise about its own shaft
(B) less than one turn clockwise about its own shaft
(C) more than one turn counter-clockwise about its own shaft
(D) less than one turn counter-clockwise about its own shaft.

72. As a ship sails away from shore it appears to go below the horizon. The best explanation for this observation is that it results from the

(A) rise and fall of tides
(B) curvature of the earth's surface
(C) refraction of light
(D) effect of gravity on moving bodies.

73. Of the following, the most important reason for lubricating moving parts of machinery is to

(A) reduce friction
(B) prevent rust formation
(C) increase inertia
(D) reduce the accumulation of dust and dirt on the parts.

74. A canvas tarpaulin measures 6 feet by 9 feet. The largest circular area that can be covered completely by this tarpaulin is a circle with a diameter of

(A) 9 feet (B) 8 feet (C) 7 feet (D) 6 feet.

75. The population of Maple Grove was 1,000 in 1976. In 1977, the population increased 40 percent but in 1978, 1979, and 1980, the population decreased 20 percent, 10 percent, and 25 percent respectively. (For each year the percentage change in population is based upon a comparison with the preceding year). At the end of this period the population was most nearly

(A) 900 (B) 850 (C) 800 (D) 750.

Questions numbered 76 through 78 are to be answered on the basis of the diagram that appears below. In this diagram, pulley A and pulley B are both firmly attached to shaft C so that both pulleys and the shaft can turn only as a single unit. The radius of pulley A is 4 inches and that of pulley B is 1 inch.

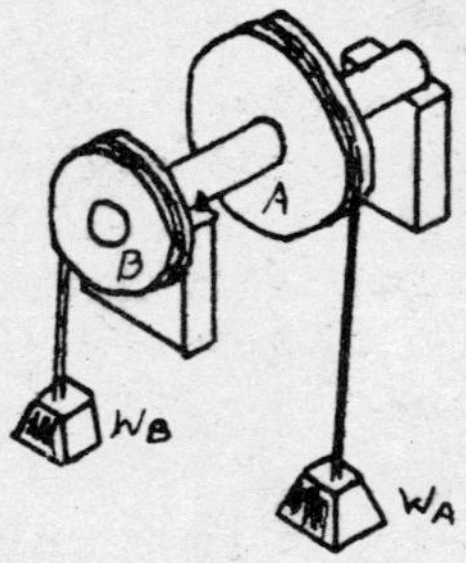

76. If the weight WA weighs 20 lbs., the system will be in balance if the weight WB weighs

(A) 5 lbs. (B) 10 lbs. (C) 20 lbs. (D) 80 lbs.

77. If the rope on pulley A is pulled downward so that it unwinds, the rope on pulley B will

(A) wind slower
(B) wind faster
(C) unwind slower
(D) unwind faster.

78. When pulley A makes one complete revolution, pulley B will make

(A) one quarter of a revolution
(B) one revolution
(C) four revolutions
(D) a number of revolutions which cannot be determined from the information given.

79. For combustion to take place, the atmosphere must contain at least 12 percent oxygen. The maximum percent of oxygen in an atmosphere which can support combustion is

(A) 21 percent
(B) 25 percent
(C) 50 percent
(D) 100 percent.

Questions numbered 80 through 82 refer to the diagram that appears below:

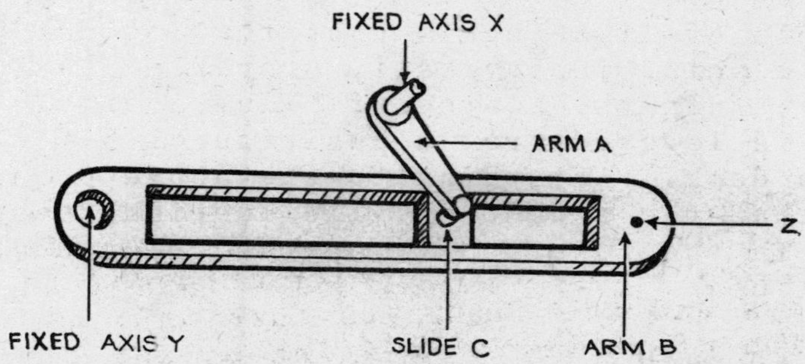

80. If arm A should rotate in a circle, arm B will most likely

(A) rotate in a larger circle
(B) move back and forth in an arc
(C) follow an unpredictable eccentric motion
(D) drop to the horizontal and lock.

81. Points X, Y and Z will be in a straight line

(A) only when slide C is at the top of the slot
(B) only when slide C is at the bottom of the slot
(C) only when slide C is at the mid-point of the slot
(D) when slide C is at either the top or the bottom of the slot.

82. If arm A should rotate at a constant speed, arm B will most likely move

(A) at an increasing speed
(B) at a decreasing speed
(C) at a constant speed
(D) at varying speeds.

83. An intense fire develops in a room in which carbon dioxide cylinders are stored. The principal hazard in this situation is that

(A) the CO_2 may catch fire
(B) toxic fumes may be released
(C) the cylinders may explode
(D) released CO_2 may intensify the fire.

84. At a fire involving the roof of a 5-story building, the firefighters trained their hose stream on the fire from a vacant lot across the street, aiming the stream at a point about 15 feet above the roof. In this situation, water in the stream would be traveling at the greatest speed

(A) as it leaves the hose nozzle
(B) at a point midway between the ground and the roof
(C) at the maximum height of the stream
(D) as it drops on the roof.

85. A principle of lighting is that the intensity of illumination at a point is inversely proportional to the square of the distance from the source of illumination. Assume that a pulley lamp is lowered from a position of 6 feet to one of 3 feet above a desk. According to the above principle, we would expect that the amount of illumination reaching the desk from the lamp in the lower position, as compared to the higher position, will be

(A) half as much
(B) twice as much
(C) four times as much
(D) nine times as much.

86. A firefighter searching an area after an explosion comes upon a man with a leg injury, bleeding severely and in great pain. Under these circumstances, the first action the firefighter should take is to attempt to

(A) immobilize the man's leg
(B) cover the man with a blanket or coat
(C) stop the bleeding
(D) make the man comfortable.

87. "When standpipes are required in a structure, sufficient risers must be installed so that no point on the floor is more than 120 feet from a riser."

The diagram which gives the maximum area which can be covered by one riser is

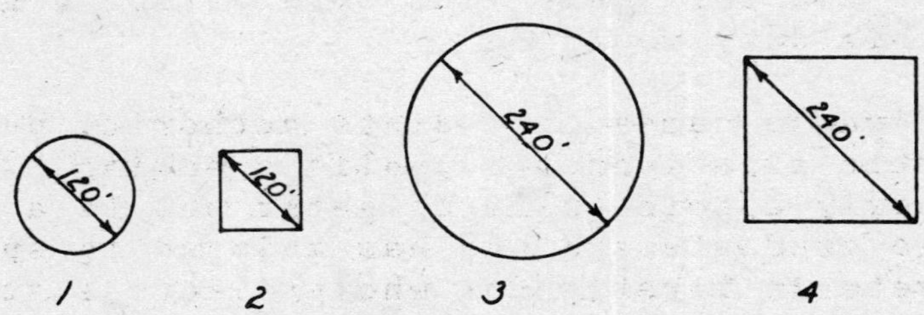

(A) 1 (B) 2 (C) 3 (D) 4.

88. Most fire departments advocate that firefighters become involved in community activities, like coaching or sponsoring a Little League team. This course of action is likely to

(A) produce expert Little League coaching
(B) lead to better understanding of the activities of the Fire Department
(C) encourage the community members to assist in fighting fires when needed
(D) enlist community support of the firefighters' quest for higher salaries.

89. There is a common bond among all firefighters functioning throughout the United States and even in some other parts of the world. There are many nationwide organizations of firefighters and it is not uncommon for them to drop in at the local fire department when they are visiting other cities. The most likely reason for this is the

(A) similarity of techniques, equipment, and problems from one fire department to another
(B) need for firefighters to be strongly organized in order to achieve equitable pay
(C) extreme dangers involved in fighting fires
(D) usual competitiveness of police and fire departments.

90. A firefighter stationed in a firehouse is approached by an irate citizen who declares that firefighters are well paid but spend most of their time lounging around the firehouse doing nothing. It would be most appropriate for the firefighter to respond by saying

(A) "You had better forward your complaint to the Mayor."
(B) "If you did my job for one week you would change your opinion."
(C) "How hard do you work for a living?"
(D) "The job of a firefighter is to always be ready to respond to an alarm."

91. Firefighters often visit schools to give talks on fire prevention to the children. The type of firefighter best suited for this duty would be

(A) one whose demeanor depicts authority but who has little experience in public speaking
(B) a newly appointed firefighter who is a recent college graduate and who has majored in speech
(C) a veteran firefighter who projects a strong image and who has addressed audiences in the past
(D) the firefighter who has achieved the highest rating on the eligible list for Firefighter.

92. During an emergency situation, a firefighter may be the recipient of an order from his or her superior that is delivered in an abrupt manner. The recipient of this type of an order should assume that

(A) he or she is in disfavor with the superior officer
(B) the superior officer feels that he or she will not respond properly to any other type of order
(C) he or she is about to be brought up on charges
(D) emergency situations call for this type of order.

93. Studies of accidents have indicated that more accidents occur at the end of a tour of duty than at its beginning. The most likely reason for this is

(A) carelessness on the part of the victims
(B) improper training in accident prevention
(C) lack of interest in safety
(D) mental and physical fatigue on the part of firefighters.

94. Firefighters can best avoid accidents while doing heavy lifting by

(A) keeping the knees bent and the back straight
(B) keeping both knees and the back straight
(C) using the arms only
(D) using the method best suited to the individual.

95. The <u>least</u> likely reason for including an accident prevention program in the training program for newly appointed firefighters is the

(A) nature of the work being performed
(B) number of reported accidents
(C) number of levels of command
(D) time lost as a result of accidents.

96. A training officer assigned to teach safety to firefighters can best gain their cooperation by

(A) exercising the authority consistent with his or her rank
(B) waiting until a serious accident occurs
(C) making a distinct effort to become "one of the boys" during the training session
(D) indicating to the trainees how an accident can be costly both to the individual and to the Department.

97. A veteran firefighter observes a newly appointed firefighter performing a maintenance task in an unsafe manner. It would be most appropriate for the veteran firefighter to

(A) tell their superior officer of the indiscretion at once
(B) illustrate the proper manner to perform this task to the rookie firefighter
(C) not do anything because this is not an assigned responsibility
(D) tell another rookie firefighter to demonstrate the proper method because it is likely that there is a common bond between them.

98. A well executed accident report is extremely valuable principally because it

(A) isolates the cause of the accident so that future similar accidents can be avoided
(B) forms the basis of statistical studies
(C) complies with existing regulations
(D) informs everybody an accident has taken place.

99. It is essential that accident reports be submitted as soon after the accident has occurred as is practical. The most important reason for this course of action is

(A) witnesses tend to change their stories as time passes
(B) that accident reports should be considered trivial and therefore they should be gotten out of the way
(C) the details of the report will still be fresh in the mind of the reporter
(D) all reports should be submitted promptly even if they are incomplete.

100. Following an accident to a firefighter on the job it is normal procedure for a Fire Department to spend considerable effort in an attempt to identify the cause of the accident. The most important reason for this procedure is

(A) no accident report is complete without the cause of the accident
(B) for statistical purposes
(C) so that the guilty firefighter may be disciplined
(D) that future similar accidents may be avoided.

SAMPLE EXAMINATION III - ANSWERS

1.	D	26.	C	51.	C	76.	D
2.	A	27.	C	52.	A	77.	A
3.	D	28.	A	53.	B	78.	B
4.	B	29.	B	54.	A	79.	D
5.	D	30.	D	55.	D	80.	B
6.	C	31.	B	56.	C	81.	D
7.	C	32.	B	57.	D	82.	D
8.	C	33.	B	58.	A	83.	C
9.	B	34.	D	59.	C	84.	A
10.	C	35.	B	60.	D	85.	C
11.	A	36.	A	61.	C	86.	C
12.	C	37.	D	62.	D	87.	C
13.	B	38.	C	63.	B	88.	B
14.	D	39.	A	64.	B	89.	A
15.	B	40.	A	65.	D	90.	D
16.	B	41.	C	66.	D	91.	C
17.	C	42.	B	67.	C	92.	D
18.	B	43.	C	68.	C	93.	D
19.	C	44.	C	69.	D	94.	A
20.	B	45.	A	70.	B	95.	C
21.	D	46.	B	71.	A	96.	D
22.	D	47.	C	72.	B	97.	B
23.	D	48.	A	73.	A	98.	A
24.	C	49.	A	74.	D	99.	C
25.	C	50.	C	75.	D	100.	D

FINAL PRACTICE EXAMINATION FOR FIREFIGHTER

Time allowed 3½ hours

Firefighters must possess the ability to readily recall the locations of doors, windows and other means of exit from a burning building. This knowledge is important for their own safety and the safety of the fire victims who they may be called upon to evacuate.

You will be given five minutes to study and attempt to memorize the floor plan of an apartment located on the fourth floor of an apartment house. When the five minute period is completed you are to commence answering the questions on the examination. The first ten questions are based upon the floor plan. You may not refer back to the floor plan while answering these questions.

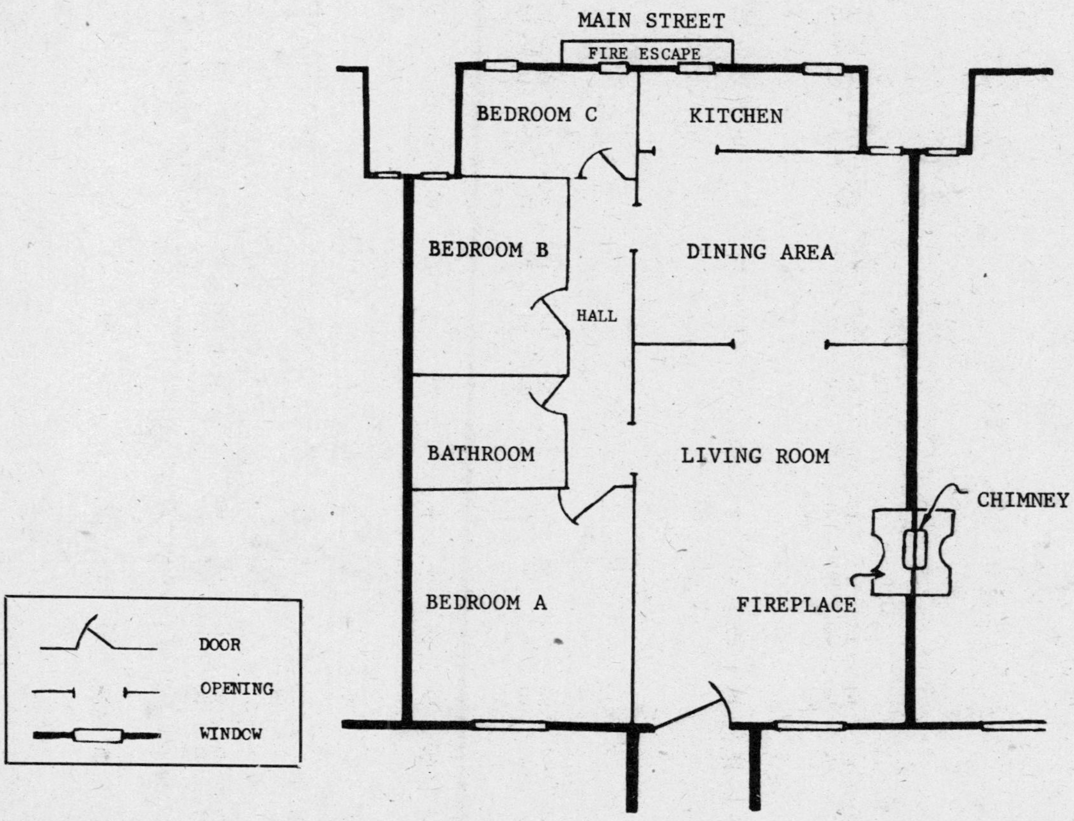

FINAL PRACTICE EXAMINATION FOR FIREFIGHTER

Time allowed - 3½ hours

Directions: The first ten questions on this examination evaluate your ability to recall the details of the floor plan you have just studied. Each question contains four choices. For each question you are to indicate the choice which you think is proper next to the appropriate number on the answer sheet. Do not refer back to the floor plan while answering these questions.

1. Which room does not have doors that can be closed?

(A) bedroom A
(B) living room
(C) dining area
(D) none of these.

2. Which room is the farthest distance from the bathroom?

(A) bedroom C
(B) living room
(C) dining area
(D) kitchen.

3. In the event of a fire in the living room, firefighters entering from the fire escape should bring a hose in through

(A) the kitchen window
(B) the hall
(C) the window of bedroom B
(D) any one of the above.

4. Firefighters would be acting properly if they checked for a fire in the apartment next door while they were extinguishing a fire in this apartment in

(A) the kitchen
(B) bedroom C
(C) the hall
(D) the chimney above the fireplace.

5. In the event of a rescue of a person in bedroom B by a firefighter during a fire in bedroom C, the best way of escape would be through the

(A) window of bedroom B
(B) hall and living room
(C) kitchen to the fire escape
(D) hall to dining area window.

6. Which room contains only one means of escape?

(A) the bathroom (C) bedroom B
(B) the living room (D) none of the above.

7. In the event that the hall was full of fire and dense smoke, a ladder would have to be used to remove a trapped person in

(A) the dining room (C) the living room
(B) the kitchen (D) bedroom B.

8. Which of the following rooms has four means of escape?

(A) bedroom A (C) kitchen
(B) dining area (D) none of them.

9. Which room does not lead directly into the hall?

(A) the bathroom (C) the kitchen
(B) the living room (D) bedroom A.

10. The most direct route from bedroom B to the fire escape would be through

(A) hall and bedroom C
(B) hall, dining area, and kitchen
(C) hall, living room, dining area, and kitchen
(D) none of the above.

Answer questions 11 through 15 on the basis of the passage below:

Most people are not aware of the services firefighters perform until they have been victimized by a fire. The public seems to have gained the impression, for the most part, that firefighters spend most of their time lounging around in a firehouse. Firefighters are able to change this image to a certain degree by courteously greeting visitors to the firehouse, by their appearance and by their deportment while in public view. Ninety percent of the rescues made by fire departments take place at small fires which do not receive the publicity that spectacular large fires do. The public rarely learns about the services firefighters perform at small fires and the rescues these entail, because fire departments do not publicize the brave acts of firefighters at routine fires.

11. What do firefighters do when they are not fighting fires?

(A) lounge around firehouses
(B) work on public relations projects

(C) repair equipment
(D) it is not indicated in this passage.

12. This passage indicates that the image of the Fire Department should be maintained by the

(A) members of the Department
(B) press and other media
(C) public it serves
(D) victims of fires.

13. The largest number of rescues is made at

(A) multiple alarm fires
(B) emergencies where there are no fires
(C) small routine fires
(D) large spectacular fires.

14. The public does not learn about rescues accomplished by firefighters at routine fires because

(A) information concerning fires should be kept confidential
(B) information concerning these fires is not very often released to the press
(C) they very often occur late at night
(D) the press is not interested in covering routine fires.

15. The most appropriate title for this passage would be

(A) Observing the Operation of a Fire Department
(B) Fire Prevention Week
(C) Improving the Public's Image of the Fire Department
(D) Bravery of Firefighters.

Answer questions 16 through 21 on the basis of the passage below:

Water and ventilation are required for effective fire fighting. Most fires can be extinguished by hosing water on them and by making it possible for gases to escape. When material which is afire is drenched with cooling water it is no longer able to produce gases which will burn. In an unventilated room hot gases can sufficiently raise the temperature of the room so that it will burst into flame. This can occur even though the room is far removed from the fire itself. Therefore, firefighters create ventilation by chopping holes in roofs and smashing windows so that gases will be able to escape.

16. Material which is burning will not give off hot gases when

(A) it is permitted to burn itself out
(B) it is exposed to the atmosphere
(C) it is cooled with water
(D) chemicals are sprayed over it.

17. A room will begin to burn when

(A) a draft is created
(B) an explosion takes place
(C) sparks are given off
(D) the temperature of the room is sufficiently raised.

18. It is possible for a room to burst into flames even though it is

(A) far removed from the actual fire
(B) saturated with water
(C) ventilated well
(D) damp and cold.

19. Very often firefighters will smash windows and chop holes in roofs. Their purpose in doing this is to

(A) escape from a fire
(B) provide access for victims
(C) ventilate the premises
(D) throw out burning materials.

20. Proper ventilation is an important firefighting technique because it

(A) provides an outlet for trapped smoke and gases
(B) causes extinguishment of flames by cooling them
(C) enables firefighters to breathe easily
(D) provides accessibility for the hoses which are being used to extinguish the flame.

21. Hot gases provide the most danger when they appear in a room that is

(A) big
(B) closed
(C) damp
(D) cold.

Answer questions 22 through 25 on the basis of the passage below:

Very often during a fire in an occupied apartment house the fire escapes become overcrowded. Firefighters are then required to attach a portable ladder to the first level of the fire escape and attach it to the opposite side of the drop ladder. If this still proves to be insufficient an additional ladder is raised from the ground to the second level and if the fire escape is located in the rear of the building, a "gooseneck" ladder that hooks over the roof can also be used. Firefighters can then assist some building occupants to go from the fire escape to the roof instead of to the ground.

22. Firefighters may use portable ladders and raise them to fire escapes so that

(A) firefighters are able to reach the roof from the exterior of the building
(B) occupants are able to reach a higher level on the fire escape
(C) firefighters are able to enter the buildings through windows
(D) occupants are able to leave fire escapes more quickly.

23. Assume that all of the ladders described in the passage are utilized. How many ways will occupants be able to reach the ground directly by ladder from the fire escape at the first level?

(A) 1
(B) 2
(C) 3
(D) 4.

24. A "gooseneck" ladder may be effectively utilized

(A) opposite the drop ladder
(B) from the top level
(C) from the first level
(D) from the second level.

25. This passage is mainly concerned with

(A) ways to deal with overcrowded fire escapes
(B) the use of portable ladders
(C) the rescue of occupants from apartments
(D) the use of the roof for occupants to escape fires.

Answer questions 26 through 30 on the basis of the passage below:

Very often fire travels inside the partitions of a burning building. Partitions are often supported by wooden studs. The fire is able to travel between the spaces separating the studs. Thus the flames are able to travel from the bottom floors to the top floors through the partitions. Falling sparks from the upper part of a partition may start a fire at the bottom of the partition. A fire that is spreading may be recognized by blistering or discolored paint or wallpaper, or partitions that feel hot to the touch. If any or all of these signs are present the partition must be opened to locate the fire. A cobweb present inside a partition is a sure sign that the fire has not spread through the partition.

26. Fires are able to spread easily through partitions because

(A) of the presence of spaces between the studs
(B) fires are able to burn anywhere
(C) partitions are constructed from readily combustible materials
(D) partitions are painted or covered with wallpaper.

27. The presence of cobwebs inside a partition are an indication that the fire has not spread through the partition because

(A) cobwebs do not burn
(B) fire is able to destroy cobwebs
(C) spiders do not build cobwebs near combustible materials
(D) cobwebs cover the spaces between studs.

28. In the event that a firefighter observes paint blistering on a partition he should immediately

(A) water down the partition
(B) go to the other rooms to see if partitions are also blistering
(C) make a hole in the partition
(D) enclose the partition by closing the doors and windows.

29. It is possible to tell if a fire is spreading within a partition by the

(A) partition's temperature
(B) smoke's color
(C) plaster's age
(D) space between the studs.

30. This passage is mainly concerned with

(A) how partitions permit fire to spread
(B) how cobwebs prevent the spreading of fires
(C) how partitions are constructed
(D) the prevention of spreading of fires.

Answer questions 31 through 36 on the basis of the following passage:

When a search is being conducted it is considered proper to first rescue victims who can be seen or heard, or those whose exact locations are known. Since some victims may be located beneath rubble the search must be properly organized. Searches which are not properly organized could result in injury or even death. The most accepted method of conducting a proper search is to start from the outer edges and to work toward the center of an area. At times a victim may be located by calling out or by tapping a pipe. It would be proper for the rescue workers to first call out and then remain silent for a period of time to await a response from a trapped victim.

31. This passage is mainly concerned with the fact that during searches

(A) firefighters should alternately call and listen
(B) trapped victims can always be heard
(C) it is necessary to be well organized
(D) victims who are buried should be located first.

32. It is accepted procedure to rescue first victims who are

(A) unconscious
(B) trapped
(C) able to be readily heard and seen
(D) hurt.

33. During the search for buried victims it is of utmost importance for firefighters to

(A) maintain periods of silence
(B) constantly keep on the move
(C) quickly search rubble piles
(D) keep clear of rubbish piles.

34. The most effective way to search an area is

(A) around the outer edges
(B) from the center of the area to the outer area
(C) from one corner to the opposite corner
(D) from the outer edge to the center of the area.

35. Disorganized movement on the part of a rescue worker can result in

(A) panic and confusion
(B) destruction of property
(C) a waste of time
(D) injury or death.

36. The method of tapping on pipes is an effective way to locate victims because

(A) it enables firefighters to use Morse code
(B) sound travels through a pipe

(C) it enables firefighters to signal one another
(D) pipes always remain free of rubble.

Answer questions 37 through 41 on the basis of the passage below:

People who intentionally set fires are called arsonists. They do not resemble criminals in appearance. However their actions very often result in tremendous property losses and frequently in loss of life. Arsonists set fires for many different reasons. A shopkeeper on the verge of bankruptcy may set a fire to collect insurance benefits. Another arsonist may seek revenge against a person for a real or imagined wrongdoing. Many arsonists take delight in seeing the fire burn and in observing the firefighters at work. At times this type of arsonist has been known to assist the firefighters in combating the fire.

37. This passage states that arsonists

(A) have a tendency to return to the scene of their crime
(B) perform their crimes mostly at night
(C) do not look like ordinary criminals
(D) always remove their fingerprints.

38. An arsonist may be defined as an individual who

(A) sets a fire with intent
(B) takes pleasure watching fires
(C) is a vengeful person
(D) is in need of money.

39. Some arsonists assist firefighters in combating fires. They do this because they

(A) have guilt feelings
(B) enjoy the excitement
(C) are compensated for their efforts
(D) are concerned for the safety of others.

40. At times a shopkeeper may become an arsonist in order to

(A) commit suicide
(B) collect insurance benefits
(C) conceal a crime
(D) increase prices.

41. This passage conveys the message that arsonists

(A) desire to become firefighters
(B) do not break existing laws

(C) suffer from a mental illness
(D) cannot be grouped together.

Answer questions 42 through 45 on the basis of the passage below:

On occasions when a fire truck is being backed into the firehouse, all firefighters are required to remain outside the building. Those who are assigned to stop oncoming traffic should face towards that traffic so that they are in a position to alert the driver if necessary. Other firefighters should be stationed on the sidewalk to guide the driver. The driver should immediately bring the truck to a complete halt if he is signaled to do so by any of the firefighters. After the truck is completely in the firehouse, only the officer in command may contact central headquarters so that the company may be placed back in service. The officer then should order the entrance doors closed.

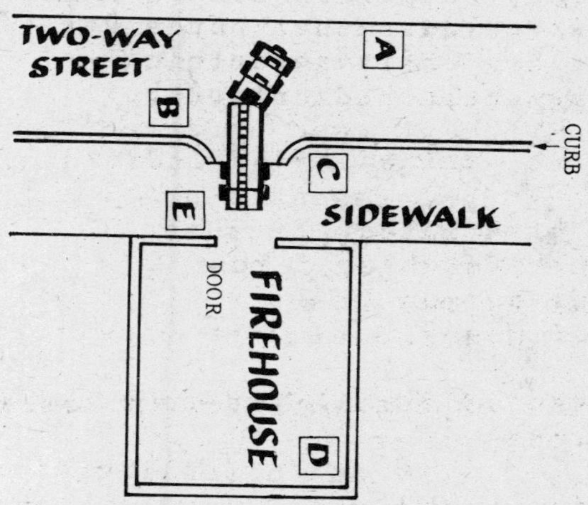

42. According to the regulations, which firefighter is not needed when the truck is being backed into the firehouse?

(A) firefighter A
(B) firefighter B
(C) firefighter C
(D) firefighter D.

43. The firefighters assigned to the halting of vehicles are

(A) firefighters A and C
(B) firefighters C and E
(C) firefighters A and B
(D) firefighters B and E.

44. The truck can be ordered to stop by

(A) only firefighters A and B
(B) only firefighters C and E
(C) only firefighter D
(D) any one and all of the firefighters.

45. Central headquarters is to be notified that the company is prepared to be put back into service when

(A) the truck has returned from the fire
(B) the truck has been parked in the firehouse
(C) the firehouse doors have been closed
(D) all of the firefighters are in the firehouse.

46. A firefighter may be required to administer first aid to people who are injured. It is of utmost importance that the firefighter be cognizant of certain conditions while performing this function. Of the following, which is the one condition he should <u>not</u> be concerned with?

(A) are the injured person's legs twisted in a manner which may indicate that bones are broken
(B) is the injured party breathing
(C) is there an open medication bottle nearby
(D) is it an apparent case of attempted suicide.

47. Hanging mirrors can present a problem to the firefighter while fighting a fire because

(A) a firefighter may see a reflection of a fire in a hanging mirror and aim the hose at it instead of at the actual fire
(B) mirrors are expensive and to break one would antagonize the owner
(C) mirrors are heavy and a falling mirror can cause injury to a firefighter
(D) the reflection of a fire from a mirror makes it burn hotter.

48. Very often pet dogs are able to save the lives of their masters by awakening them before a fire has gotten out of control. Dogs are usually able to detect fires before their masters because

(A) dogs are more used to fires than people
(B) dogs' sensitivity to smoke is greater than people's
(C) dogs always recognize danger
(D) fires make dogs thirsty.

49. It is <u>least</u> important for a firefighter assigned to inspect a building to check the

(A) number and location of fire extinguishers
(B) areas utilized to store combustible materials
(C) main electric control switch
(D) public washrooms.

50. Of the following fires, which one would be most likely to have been set intentionally

(A) fire in a frying pan
(B) fire in a mailbox
(C) fire in a child's bedroom
(D) fire in an attic.

51. A firefighter in the act of chopping a hole in a roof to let smoke escape would be in <u>least</u> danger from the smoke if the wind was blowing

(A) from his left
(B) toward his face
(C) toward his back
(D) from his right.

52. The hands of firefighters are injured more in the summer than they are in the winter. The most likely reason for this is that

(A) objects tend to retain their heat more in the summer than they do in the winter
(B) firefighters' hands tend to sweat in the summer
(C) fires tend to burn hotter in the summer
(D) firefighters tend to remove their protective gloves while working during the summer.

53. Standard operating procedures dictate that firefighters approach closely the fire they are fighting while attempting to extinguish it. This is done mainly to

(A) ascertain what started the fire
(B) achieve a degree of accuracy in aiming their hoses
(C) use as short a length of hose as possible
(D) avoid the thick smoke.

54. Of the following pieces of equipment, the one that would be of least help to a firefighter attempting to extinguish a fire at the scene of an automobile accident would be a

(A) ladder which is portable
(B) fire extinguisher which is portable
(C) stretcher
(D) length of hose.

55. Very often firefighters cross a burning room on their hands and knees because

(A) they are less likely to fall
(B) they will thus avoid flying sparks
(C) fire will not spread across a floor
(D) there is less smoke at floor level.

56. One of the functions of a firefighter is to inspect buildings to see that fire safety laws are enforced. It is best to conduct these inspections unannounced because

(A) of the waste of time involved if a building owner breaks an appointment
(B) building owners may try to conceal fire hazards
(C) time will be saved
(D) owners will be inconvenienced if firefighters are delayed.

57. Firefighters have reason to believe that the three fires in an unused school house that they have had to extinguish in the past month have been set deliberately. Therefore the Fire Department has decided to take pictures of the crowd observing them fighting any future fires in the building. This course of action would be helpful because

(A) it will further the Fire Department's public relations program
(B) it will be used to convince the city to demolish the building
(C) it will discourage the crowd from getting too close
(D) arsonists tend to return to observe fires that they have set being extinguished.

58. A firefighter would be exposed to the greatest danger in the act of extinguishing a fire in a

(A) florist shop
(B) supermarket
(C) paint store
(D) bank.

59. Firefighters responding to an alarm find a badly beaten unconscious man lying next to the alarm box. It would be appropriate for the firefighters to do all but which one of the following?

(A) summon the police
(B) give first aid to the man
(C) summon an ambulance
(D) question the homeowners in the vicinity.

60. During the times when firefighters are not responding to alarms they busy themselves checking the equipment and vehicles. The most likely reason for them to engage in this activity is to

(A) be busy when the public visits their firehouse
(B) provide free time to socialize
(C) promote neighborhood pride in the appearance of the firehouse
(D) assure that their equipment and vehicles are in good working order.

61. Shopkeepers will be most likely to abide by fire safety rules if they

(A) comprehend the reasons for the rules
(B) are not too busy
(C) fear the penalties that will be assessed if they do not abide by the rules
(D) have just opened a new place of business.

62. In a large public building the exit doors should properly

(A) open outward
(B) slide open
(C) revolve
(D) open inward.

63. To delay the spread of a fire from one room to another you should

(A) punch a hole in a wall between the two rooms
(B) keep all windows in both rooms closed
(C) remove all the furniture from both rooms
(D) keep all doors closed between the two rooms.

64. The need for rescue work would be greatest in a one story building which is aflame and houses

(A) eighty high school students
(B) eighty retail customers
(C) eighty residents of a nursing home
(D) eighty office workers.

65. Fires in industrial plants are more likely to cause extensive damage if they occur at night when the plants are not in use than if they occur during the day because

(A) it takes firefighters longer to respond to a fire at night
(B) fires are recognized sooner in an occupied building
(C) arsonists usually function at night
(D) a fire tends to smolder in cool temperatures.

66. As the firefighters are preparing to leave the scene of a fire a reporter stops one of them and questions him about the details of the fire. The firefighter would be acting properly if he responded by telling the reporter

(A) that firefighters are not permitted to speak to reporters
(B) that he should speak to his or her supervisor
(C) to call and make an appointment for an interview at the firehouse
(D) that other reporters at the fire already have the details and the reporter should speak to them.

67. While fighting a fire, a firefighter becomes trapped in a third-floor loft when the stairs catch fire. There is no fire escape. In this instance it would be best to

(A) call for a portable ladder to escape through a window
(B) wait for help to arrive
(C) go to the nearest window and jump
(D) go up the stairs to the next floor and await a rescue attempt.

The duties of a firefighter very often put him into close contact with the public. On occasion the firehouse may be visited by groups of citizens and a firefighter may be assigned to inspect buildings for violations of fire safety laws. A firefighter will often come into contact with the public while fighting fires. Firefighters are public servants and they should treat the public with courtesy and respect. However, while fighting a fire their most important function is putting out the fire.

68. A firefighter arriving at the scene of a fire encounters an angry woman who directs her ire at the firefighter saying that she had sounded the alarm thirty minutes ago and much unnecessary damage has been caused by the delay. The firefighter knows that just six minutes have elapsed since the alarm first came in. In this instance it would be best to

(A) request of her the exact time she sounded the alarm and indicate that it was only six minutes ago
(B) offer her an explanation of just how firefighters respond to an alarm
(C) state that time seems to pass slowly when people are concerned about any matter
(D) not respond but rather attend to firefighting duties without delay.

69. A firefighter assigned to inspection duties comes across some trash piled on a stair landing. This condition is not only dangerous but it is in clear violation of a fire safety law. When this violation is called to the attention of the owner of the building, he states that the last inspector overlooked the condition because people can use the elevator instead of the stairs. It would be most proper for the fire-

fighter to <u>first</u>

(A) overlook the violation because firefighters must co-operate with each other
(B) insist that the owner remove the trash at once
(C) request the name of the previous inspector from the owner
(D) ask the owner for the reason that he stores trash on the stair landing.

70. A firefighter, on fire inspection duty, enters a restaurant which is very busy. Ordinarily the owner accompanies the firefighter during the inspection but on this day the owner states that he is short of help and does not have the time to do so. He requests that the firefighter return during a less busy period when he will be able to accompany the firefighter during the inspection. In this instance it would be most proper for the firefighter to

(A) agree to return at a later date after informing the owner that the inspection will be harsh
(B) proceed with the inspection without the owner because it would be a good time to detect violations
(C) inform the owner that he will be reported to a superior officer
(D) stick around the premises until the owner has time.

71. Firefighters at the scene of a fire have just about succeeded in putting out the fire. Before it has been completely extinguished a couple who live on the top floor request permission to enter their apartment to secure some of their belongings. It would be most proper for the firefighter who receives this request to

(A) permit the couple to enter their apartment since the fire is almost out
(B) suggest that the couple give him a written list and he will obtain himself whatever the couple is in need of
(C) permit one, not both, to return to the apartment
(D) inform them that they will be permitted to enter only when the building is considered completely safe.

72. A man excitedly enters a firehouse and tells the firefighters on duty that children are throwing large rocks at passing cars from an overpass. In this instance it would be best for the firefighters to

(A) accompany the man to the overpass and stop the children
(B) report the situation at once to their superior officer
(C) drive the man to the nearest police station
(D) inform the man that stopping the children is not one of their duties.

73. While extinguishing a dangerous fire, a firefighter receives a complaint from a woman that she has been sworn at by another firefighter for being in the way. It would be most appropriate for the firefighter to

(A) promptly apologize and to explain in detail why the other firefighter acted in that way
(B) do or say nothing because the woman was wrong
(C) inform the woman that she should report the incident to the supervisor after the fire has been completely extinguished
(D) inform the other firefighter that he or she should apologize otherwise he/she will be in trouble.

74. Twenty-two firefighters are assigned to a firehouse. Twenty have indicated a desire to purchase a color TV set, the other two do not want to contribute their shares at this time because they cannot afford to. The most appropriate manner to handle this situation is to

(A) suggest to the two firefighters that they try to earn some extra money in order to pay their shares
(B) cancel plans to buy a new set at this time
(C) suggest to the two that they try to borrow enough money to cover their shares
(D) purchase the set at this time and permit the two to contribute towards it whenever they are able to.

75. In the act of inspecting a small store during business hours, a firefighter observes an exit door which is padlocked. Fire safety laws state that all exit doors must remain unlocked while an establishment is open for business. The owner says that burglars had once entered the store through the exit door and it is now padlocked with a key hanging on a nail as protection against future burglaries. It would be most proper in this instance for the firefighter to

(A) give an explanation of the requirements of the appropriate law and direct that the padlock be removed during business hours
(B) have all of the employees demonstrate how quickly they can open the door
(C) take no action because the key is readily accessible
(D) direct the owner to supply each employee with an individual key.

76. A fire truck is stationed at a fireworks display as a precautionary measure in case a fire should break out. As the fireworks begin, children commence climbing onto the truck. The firefighters should

(A) show the children around the truck and explain its operation
(B) permit two children on the truck at a time so that they may be closely observed
(C) order the children off the truck so that they will not be in the way if the truck is put into service
(D) move the truck a distance from the fireworks so that its presence will not attract the children.

77. A superior officer orders a firefighter to use a specific type of hose during a fire. The firefighter feels that it would be better to use a different type of hose. The firefighter would be responding properly to this situation if he or she

(A) used his/her type of hose when the superior was not watching
(B) requested that the superior give a reason for selecting the hose
(C) followed the order of the superior and asked for an explanation after the fire had been extinguished
(D) explained to the superior why his/her choice of hose is superior.

78. A firefighter, on his first day on the job, is the recipient of a complaint from a resident of the neighborhood that too much money is being spent on expensive firefighting equipment. A proper response to this complaint would be

(A) "Permit me to get my superior to discuss this matter with you, I am newly appointed."
(B) "This is a big city and it needs the best available equipment."
(C) "Sir, you are wrong, but you are entitled to your opinion."
(D) "You should write to the Mayor concerning your complaint."

79. An obviously intoxicated man staggers into the firehouse and asks for help to start his car which is stuck about a block away from the firehouse. In this instance it would be most appropriate for the firefighter to

(A) accompany the man to his car to see what is wrong
(B) ask the man just how much he has had to drink
(C) advise the man not to drive but to take a cab home
(D) take the man home.

80. A Scout troop visits the firehouse to inspect the equipment and to speak to the firefighters about their work. This is considered a

(A) good practice because firefighters will thus be kept busy
(B) bad practice because firefighters should be permitted to rest when they are not fighting fires
(C) good practice because it presents an opportunity to teach the Scouts about fire safety
(D) bad practice because the Scouts can do damage to the equipment.

81. During an inspection a store owner tells a firefighter that inspections are a waste of money and time and that the Fire Department does a rotten job. It would be best for the firefighter to

(A) tell the store owner to clear up his own deficiencies before complaining about the Fire Department
(B) suggest that they trade jobs for a day
(C) attempt to find out just why the store owner feels as he does
(D) tell the store owner that he has a very bad attitude.

82. After a fire company has completed extinguishing a fire in a store, the owner complains to a firefighter that another firefighter has chopped a hole in the roof, thus causing more damage than the fire itself. It would be most appropriate for the firefighter to

(A) inform the owner that insurance will cover the damage
(B) direct the owner's complaint to the firefighter who had chopped the hole
(C) explain to the store owner the reason for chopping the hole
(D) suggest that the owner lodge his complaint with the Fire Commissioner.

83. A firefighter, upon completing a lecture on fire prevention at a school, finds that the children have more questions than can be answered during the allotted time. It would be most appropriate for the firefighter to

(A) respond to as many questions as possible and attempt to arrange for another visit
(B) permit only the teacher to answer questions
(C) remain until all of the questions are answered
(D) avoid answering any questions.

84. Neighborhood children have been visiting the firehouse several times during the day. They are friendly and reasonably well behaved but their frequent presence has interfered with the firefighters' performance of their maintenance duties. The most appropriate course of action for the firefighters to take is to

(A) instruct the children not to return to the firehouse
(B) call the parents and request that they keep their children from visiting the firehouse
(C) suggest that the children visit their neighborhood police station
(D) request the children leave because the firefighters have work to do.

85. After a fire has been extinguished, a firefighter observes a newly appointed firefighter pulling down loose plaster from a ceiling without having his protective helmet on.

It would be most appropriate in this instance for the experienced firefighter to

(A) advise the superior officer concerning the rule violation
(B) say nothing to anyone
(C) explain to the new firefighter the danger involved and request that he/she wear a helmet while performing this function
(D) go to any extreme, using force if necessary, to convince the new firefighter to wear the helmet.

86. Firefighters ordinarily prepare their own meals while on duty in the firehouse. Most firefighters like to take their turn cooking meals. However, one firefighter offers to wash the dishes rather than take a turn cooking. It would be best in this instance for the other firefighters to

(A) attempt to convince the firefighter that cooking can be fun
(B) accept the firefighter's offer
(C) permit the firefighter to make sandwiches instead of hot meals during his turn
(D) permit the firefighter to refrain from both cooking and doing the dishes.

87. Every Fire Department desires to achieve the support and confidence of the community it serves. As far as the Fire Department is concerned this support and confidence can best be demonstrated by the citizenry's

(A) assistance in extinguishing large fires
(B) assistance in the maintenance of equipment
(C) support of the firefighter's wage negotiations
(D) participation in fire prevention programs and their adherence to fire protection laws.

88. An off-duty firefighter sprained an ankle while playing football. The next day the firefighter returned to work even though the ankle was quite painful. The firefighter felt that he/she was needed on the job. This course of action is con-

sidered improper mainly because the firefighter should have known that

(A) no one is indispensable
(B) medical attention was more necessary than reporting for duty
(C) because of this injury it might be impossible to perform his or her share of the work when required at a fire
(D) since the injury occurred off duty the medical expenses should not be borne by the Fire Department.

89. While fighting a fire, a firefighter searches an apartment looking for victims and comes across an unconscious woman on a bed. The firefighter would probably be of most assistance to the woman by <u>first</u>

(A) calling for a doctor
(B) giving the woman first aid until she regains her consciousness
(C) moving the woman to a safe place
(D) remaining with the woman in case she regains consciousness.

90. After a fire in an apartment has been extinguished and as you, a firefighter, are returning equipment to the fire truck, you are falsely accused by a woman of having taken twenty dollars she had hidden in her kitchen cupboard. In this situation it would be proper for you to

(A) deny the charge and to advise the woman to lodge her complaint with your superior officer
(B) offer to return the missing amount of money to get rid of her
(C) tell her she had no business trying to hide her money
(D) ask her if she thought a firefighter would risk his job for this small amount of money.

Questions 91 through 96 evaluate your knowledge of tools and the purposes for which they were intended. For each question, you are to decide which one of the four implements depicted in the boxes labeled A, B, C or D is normally used with the implement depicted in the picture on the left. You are then to indicate your selection next to the appropriate number on your answer sheet. Note: The tools are <u>not</u> drawn to scale.

91.

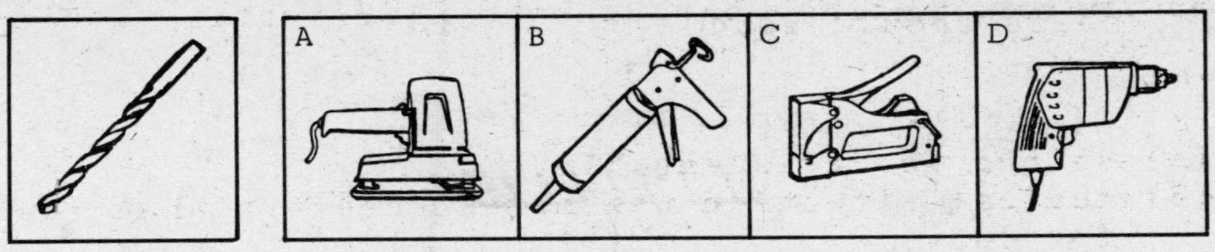

92.

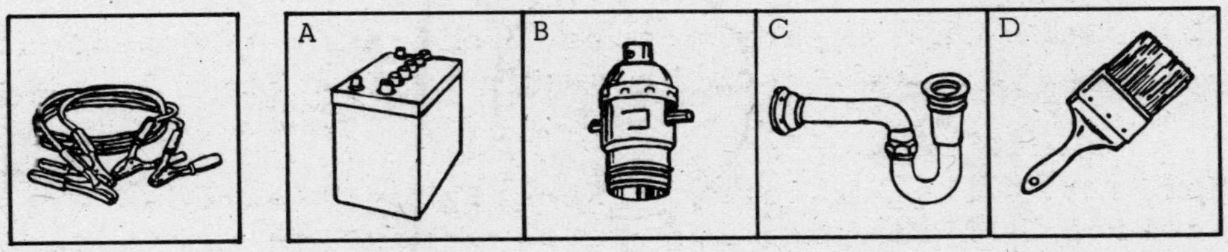

93.

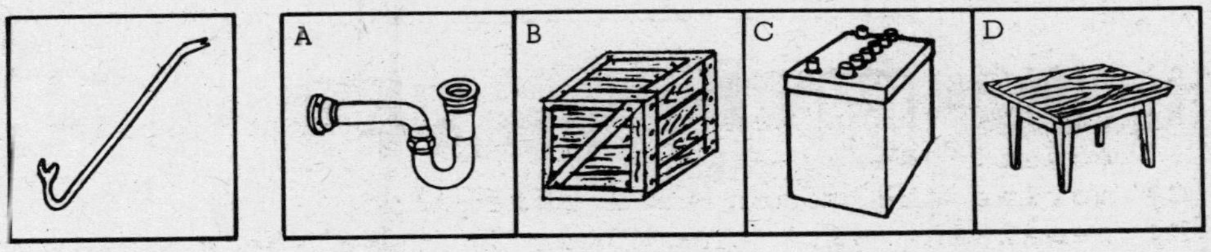

94.

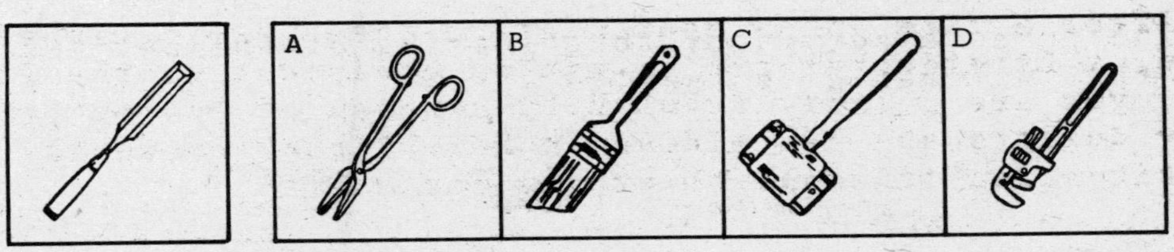

95.

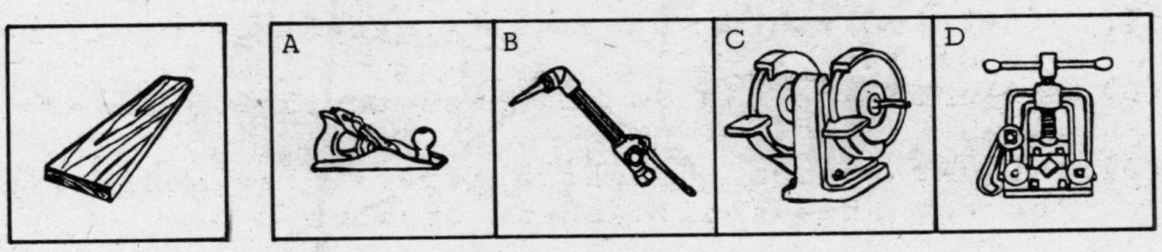

96.

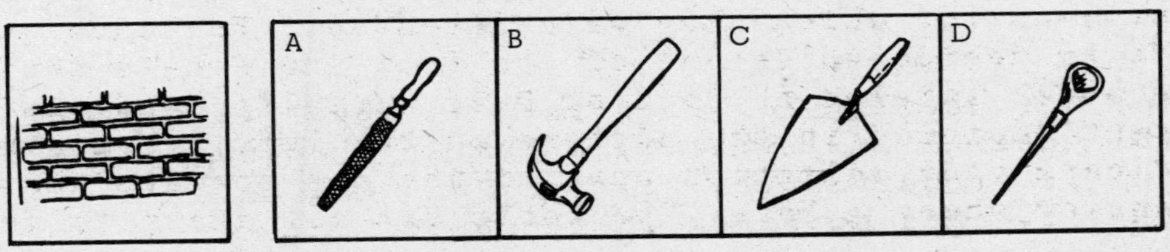

97. The scale should read

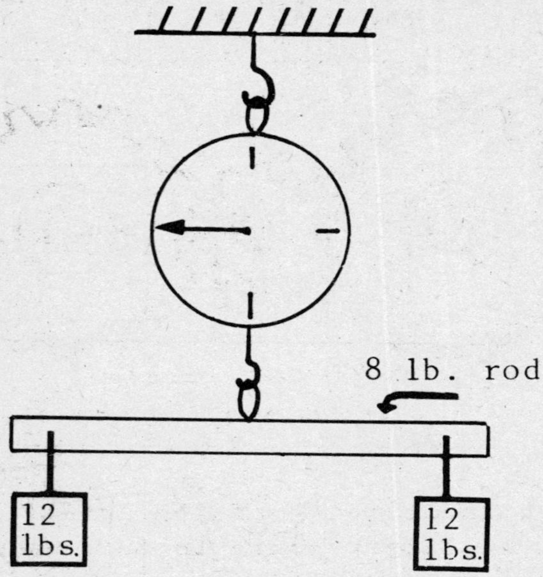

(A) zero weight (B) 12 pounds (C) 16 pounds (D) 32 pounds.

98. The purpose served by the crossing of the belt which connects the two wheels is that

(A) the wheels will then turn in opposite directions
(B) wheel A will then turn faster than wheel B
(C) it preserves the belt
(D) it takes up slack in the belt.

99. The purpose served by the small gear situated between the two larger gears is to

(A) speed up the turning of the larger gears
(B) permit the larger gears to turn in opposite directions
(C) cut down the speed of the larger gears
(D) direct the larger gears to turn in the same direction.

100. All three of the two-foot-high water containers depicted below have an area of one square foot at their base. The pressure on the bottom of the containers is

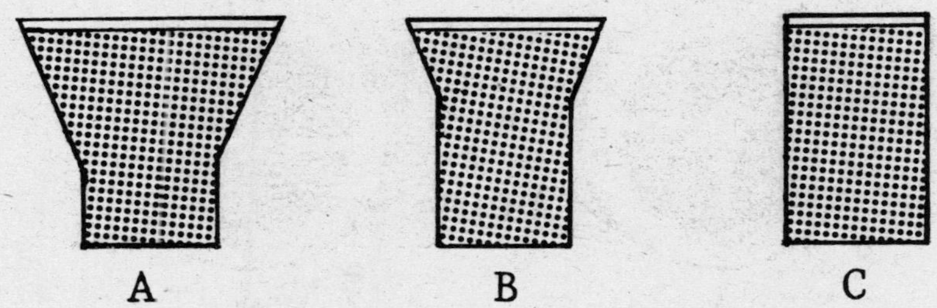

(A) the greatest in A
(B) the greatest in B
(C) the greatest in C
(D) equal in all three containers.

FINAL PRACTICE EXAMINATION - KEY ANSWERS

1.	C	26.	A	51.	C	76.	C
2.	D	27.	B	52.	D	77.	C
3.	A	28.	C	53.	B	78.	A
4.	D	29.	A	54.	A	79.	C
5.	B	30.	A	55.	D	80.	C
6.	A	31.	C	56.	B	81.	C
7.	D	32.	C	57.	D	82.	C
8.	B	33.	A	58.	C	83.	A
9.	C	34.	D	59.	D	84.	D
10.	A	35.	D	60.	D	85.	c
11.	D	36.	B	61.	A	86.	B
12.	A	37.	C	62.	A	87.	D
13.	C	38.	A	63.	D	88.	C
14.	B	39.	B	64.	C	89.	C
15.	C	40.	B	65.	B	90.	A
16.	C	41.	D	66.	B	91.	D
17.	D	42.	D	67.	A	92.	A
18.	A	43.	C	68.	D	93.	B
19.	C	44.	D	69.	B	94.	C
20.	A	45.	B	70.	B	95.	A
21.	B	46.	D	71.	D	96.	C
22.	D	47.	A	72.	B	97.	D
23.	B	48.	B	73.	C	98.	A
24.	B	49.	D	74.	D	99.	D
25.	A	50.	B	75.	A	100.	D

FINAL PRACTICE EXAMINATION - EXPLANATORY ANSWERS

1. (C) The dining area has two entrances but they are without doors.

2. (D) Of the choices given, the kitchen is farthest removed from the bathroom.

3. (A) By bringing the hose through the kitchen window and running it through the entrances in the dining area, the hose will be subjected to the smallest amount of bending possible.

4. (D) Of the choices given, this is the only area which abuts the next apartment.

5. (B) The firefighter would be exiting from the apartment going away from the fire.

6. (A) The bathroom does not have a window, only a door.

7. (D) The door leading to the hall could not be used, therefore the window would have to serve as the means of exit.

8. (B) Three entrances and one window.

9. (C) The kitchen is the only room which does not lead directly into the hall.

10. (A) The shortest amount of distance will then be covered.

11. (D) Nowhere in the passage does it say what the firefighters are doing when they are not fighting fires.

12. (A) This is stated in the third sentence of the passage.

13. (C) This is stated in the fourth sentence of the passage.

14. (B) Refer to the last sentence in the passage.

15. (C) The overall implication of the passage is that it will be up to the firefighters themselves to maintain a satisfactory image to the public.

16. (C) Refer to the third sentence in the passage.

17. (D) Refer to the fourth sentence in the passage.

18. (A) This is stated directly in the fifth sentence of the passage.

19. (C) The final sentence not only defines the term ventilation but it also states its purpose.

20. (A) Once more refer to the last sentence in the passage.

21. (B) The entire purpose of ventilation is to provide an opportunity for gases to escape.

22. (D) They provide an additional means of exit from the fire escape.

23. (B) From the regular fire escape ladder and the portable ladder attached to the opposite end of the fire escape by the firefighters.

24. (B) It is attached to the edge of the roof and reaches to only the top level of the building.

25. (A) The beginning of the passage depicts a problem which occurs quite often while fighting fires in apartment houses. The balance of the passage is devoted to the means of overcoming that problem.

26. (A) The size of the space between studs is usually dictated by building codes.

27. (B) Fire will demolish cobwebs instantly.

28. (C) To locate the fire in order to fight it. (Next to last sentence in the passage.)

29. (A) Fourth sentence of the passage. A partition which is hot to the touch indicates a fire is burning behind it.

30. (A) Almost the entire passage is devoted to just how fire spreads inside a partition.

31. (C) This is brought out in the final sentence of the passage.

32. (C) This is stated in the first sentence of the passage.

33. (A) This is implied in the final sentence of the passage. The searchers must provide the opportunity for trapped victims to be heard as they cry for help.

34. (D) As stated in the fourth sentence of the passage. Using this procedure will insure that the entire area is covered.

35. (D) Refer to the third sentence in the passage.
36. (B) If you do not remember it from your high school physics, it is stated in the next to the last sentence of the passage.

37. (C) As stated in the second sentence in the passage.

38. (A) Right in the first sentence of the passage.

39. (B) As stated in the next to the last sentence of the passage.

40. (B) Directly stated in the fifth sentence in the passage.

41. (D) This is implied from the fourth sentence on.

42. (D) The regulations do not call for a firefighter to be stationed in the firehouse when the fire truck is being backed into it.

43. (C) They are stationed in the street in proper position to halt oncoming vehicles.

44. (D) Directly stated in the fourth sentence of the passage.

45. (B) As stated in the next to the last sentence of the passage.

46. (D) The purpose of first aid is to provide immediate medical attention until professional help arrives; not to determine the reason for the disability.

47. (A) This very often occurs and it costs firefighters valuable time.

48. (B) Dogs are far more sensitive to smoke than human beings.

49. (D) It is of most importance that the inspecting firefighter determines that there are a sufficient number of fire extinguishers present and that they are located in the proper areas. The areas where combustible materials are stored and location of the main electric control switch are also very important and should be carefully inspected. Fires do not often occur in public washrooms.

50. (B) It takes a deliberate act to start a fire in a mailbox.

51. (C) The smoke would then be blown away from him.

52. (D) It always happens wherever protective gloves are used. Workers tend to remove them during warm weather.

53. (B) The shorter the distance they have to train their hoses, the more accuracy they are likely to achieve.

54. (A) There is no point in attaining a height while fighting a fire in an automobile.

55. (D) Smoke is lighter than air and therefore it rises.

56. (B) Inspections should be made when ordinary conditions exist. A prior announcement of an inspection will give an individual time to prepare for it.

57. (D) The facts in the other choices may not be assumed. Many arsonists derive a great deal of pleasure from observing fires that they have set.

58. (C) Paints and their by-products are inflammable.

59. (D) The immediate function of the firefighters would be to give first aid, to summon professional medical assistance, and then to summon police-in that order.Finding out what happened should be left to the police.

60. (D) Firefighters spend all the time necessary between answering alarms to make sure their equipment is functioning properly. Ineffective equipment prevents firefighters from functioning properly.

61. (A) All citizens tend to abide by rules and regulations that they understand including their purpose.

62. (A) This will make for easy evacuation of the building.

63. (D) The key word is <u>delay</u>, it does not mean to prevent.

64. (C) The elderly residents would probably need individual help while being evacuated.

65. (B) The sooner the fire is detected the easier it is to extinguish.

66. (B) The superior officer would be able to respond in the manner set by the Fire Department. Contact between the press and members of a Fire Department should be on a limited basis in order to maintain consistency.

67. (A) It would be improper for him to just wait and sit back waiting for help to arrive. On the other hand, the preamble of the question does not indicate that the fire has advanced to the degree that the firefighter has to jump. The preamble does indicate that the stairs are aflame so why go

back to them. The correct answer is, therefore, to call for a portable ladder.

68. (D) The function of a firefighter arriving at the scene of a fire is to extinguish it, not to promote good public relations.

69. (B) There is no way that the building owner can justify this condition; therefore, the inspecting firefighter should act in a direct manner and correct the violation without any delay.

70. (B) The most effective inspections are carried out without giving prior notice. This more or less eliminates the possibility of preparing the premises so that it can pass the inspection, then permitting violations in between inspections.

71. (D) The safety of people is primary and comes before everything else.

72. (B) Choices A and C would remove the firefighters from their primary function, availability to respond to fire alarms. Choice D would make for bad public relations. Of the choices given, choice B is the best. The superior officer can then decide to either summon the police or to send a limited force of firefighters.

73. (C) The preamble indicates that the fire is dangerous and at the point of the complaint it is still to be extinguished. After the fire has been extinguished the complaint will be the responsibility of the superior officer to hear and resolve.

74. (D) Because the overwhelming majority of the firefighters wish to purchase the set, the course of action set down in choice D is best. The courses of action stated in choices A and C are likely to be resented by the two firefighters.

75. (A) This violation creates a hazardous situation and calls for direct corrective measures.

76. (C) The presence of the children on the truck will present a dangerous situation if the truck is put into service. The function of the fire truck is to provide service, not to promote public relations.

77. (C) The function of a firefighter is to execute orders from superior officers first and to ask questions concerning them afterwards.

78. (A) The newly appointed officer is not equipped to defend the position of the Fire Department.

79. (C) Keep the intoxicated driver off the road at any cost.

80. (C) The situation in the preamble of the question is considered good. Choice C is far better than choice A because it is more objective.

81. (C) If you can determine the basis for the store owner's feelings, perhaps you can find a way to correct them.

82. (C) If the owner understands the reason for the chopping of the hole in the roof it will be easier for him to accept.

83. (A) It is very important that all valid questions be answered so that all doubts can be removed. In the event that time is limited additional time should be scheduled.

84. (D) The preamble of the question states that the children are interfering with the functions of the firefighters. Therefore, direct corrective action is called for.

85. (C) Proper corrective measures should be taken immediately before the onset of an accident.

86. (B) The officer in this choice seems to be sincere, so why press the issue? Choice D is ridiculous.

87. (D) This is the ideal as far as a Fire Department is concerned.

88. (C) A firefighter who is not physically prepared to perform his or her functions can be the cause of much harm not only to him or herself but also to fellow firefighters.

89. (C) The woman's safety comes first. Choice B is not correct because the firefighter has no assurance that the woman will regain consciousness quickly.

90. (A) This is a serious charge despite the relatively small sum involved and should be handled formally.

91. (D) The drill uses the bit to make holes.

92. (A) The cables connect batteries.

93. (B) A crowbar is used to open crates.

94. (C) A chisel is tapped by a mallet.

95. (A) A plane is used to smooth wood.

96. (C) A trowel spreads mortar in the building of a brick wall.

97. (D) The spring in the scale reflects all of the weights connected to the hook.

98. (A) That is exactly the purpose of a crossed belt between two wheels.

99. (D) A small gear installed between two larger gears will cause them to turn in the same direction.

100. (D) Water pressure is equal if the area is equal in size.

PART 3

Previous Test Questions Arranged to Help You Study Each Subject That May Be Expected on the Exam

SAMPLE ANSWER SHEET

EXAM TITLE		TODAY'S DATE
SCHOOL OR BUILDING	ROOM NO.	SEAT NO.

PRINT WITH SOFT PENCIL ONLY. Print, with pencil, your SOCIAL SECURITY NO. and the EXAM. NO. in the boxes at the tops of the columns. ONE NUMBER IN A BOX. In each column, darken (with pencil) the oval containing the number in the box at the top of the column. Only ONE OVAL in a COLUMN should be darkened. Then, using your pencil, print in the: Exam. Title, School or Building, Room No., Seat No., and Today's Date.

Follow the instructions given in the question booklet. Mark nothing ▶ but your answers in the ovals below.
SAMPLE QUESTION: When we add 5 and 3 we get: (A) 11 (B) 9 (C) 8 (D) 2. Since the answer is 8, your answer should be marked like this: (A) (B) ● (D)
WARNING: Be sure that the oval you fill in is in the row numbered the same as the question you are answering. Use a No. 2 pencil (soft pencil).
BE SURE YOUR PENCIL MARKS ARE HEAVY AND BLACK. ERASE COMPLETELY ANY ANSWER YOU WISH TO CHANGE. DO NOT make stray pencil dots, dashes or marks ANYPLACE on this SHEET.

YOUR SOCIAL SECURITY NO.								
0	0	0	0	0	0	0	0	0
1	1	1	1	1	1	1	1	1
2	2	2	2	2	2	2	2	2
3	3	3	3	3	3	3	3	3
4	4	4	4	4	4	4	4	4
5	5	5	5	5	5	5	5	5
6	6	6	6	6	6	6	6	6
7	7	7	7	7	7	7	7	7
8	8	8	8	8	8	8	8	8
9	9	9	9	9	9	9	9	9

EXAM NO.			
0	0	0	0
1	1	1	1
2	2	2	2
3	3	3	3
4	4	4	4
5	5	5	5
6	6	6	6
7	7	7	7
8	8	8	8
9	9	9	9

START HERE

1 (A) (B) (C) (D)	26 (A) (B) (C) (D)	51 (A) (B) (C) (D)	76 (A) (B) (C) (D)	101 (A) (B) (C) (D)	126 (A) (B) (C) (D)
2 (A) (B) (C) (D)	27 (A) (B) (C) (D)	52 (A) (B) (C) (D)	77 (A) (B) (C) (D)	102 (A) (B) (C) (D)	127 (A) (B) (C) (D)
3 (A) (B) (C) (D)	28 (A) (B) (C) (D)	53 (A) (B) (C) (D)	78 (A) (B) (C) (D)	103 (A) (B) (C) (D)	128 (A) (B) (C) (D)
4 (A) (B) (C) (D)	29 (A) (B) (C) (D)	54 (A) (B) (C) (D)	79 (A) (B) (C) (D)	104 (A) (B) (C) (D)	129 (A) (B) (C) (D)
5 (A) (B) (C) (D)	30 (A) (B) (C) (D)	55 (A) (B) (C) (D)	80 (A) (B) (C) (D)	105 (A) (B) (C) (D)	130 (A) (B) (C) (D)
6 (A) (B) (C) (D)	31 (A) (B) (C) (D)	56 (A) (B) (C) (D)	81 (A) (B) (C) (D)	106 (A) (B) (C) (D)	131 (A) (B) (C) (D)
7 (A) (B) (C) (D)	32 (A) (B) (C) (D)	57 (A) (B) (C) (D)	82 (A) (B) (C) (D)	107 (A) (B) (C) (D)	132 (A) (B) (C) (D)
8 (A) (B) (C) (D)	33 (A) (B) (C) (D)	58 (A) (B) (C) (D)	83 (A) (B) (C) (D)	108 (A) (B) (C) (D)	133 (A) (B) (C) (D)
9 (A) (B) (C) (D)	34 (A) (B) (C) (D)	59 (A) (B) (C) (D)	84 (A) (B) (C) (D)	109 (A) (B) (C) (D)	134 (A) (B) (C) (D)
10 (A) (B) (C) (D)	35 (A) (B) (C) (D)	60 (A) (B) (C) (D)	85 (A) (B) (C) (D)	110 (A) (B) (C) (D)	135 (A) (B) (C) (D)
11 (A) (B) (C) (D)	36 (A) (B) (C) (D)	61 (A) (B) (C) (D)	86 (A) (B) (C) (D)	111 (A) (B) (C) (D)	136 (A) (B) (C) (D)
12 (A) (B) (C) (D)	37 (A) (B) (C) (D)	62 (A) (B) (C) (D)	87 (A) (B) (C) (D)	112 (A) (B) (C) (D)	137 (A) (B) (C) (D)
13 (A) (B) (C) (D)	38 (A) (B) (C) (D)	63 (A) (B) (C) (D)	88 (A) (B) (C) (D)	113 (A) (B) (C) (D)	138 (A) (B) (C) (D)
14 (A) (B) (C) (D)	39 (A) (B) (C) (D)	64 (A) (B) (C) (D)	89 (A) (B) (C) (D)	114 (A) (B) (C) (D)	139 (A) (B) (C) (D)
15 (A) (B) (C) (D)	40 (A) (B) (C) (D)	65 (A) (B) (C) (D)	90 (A) (B) (C) (D)	115 (A) (B) (C) (D)	140 (A) (B) (C) (D)
16 (A) (B) (C) (D)	41 (A) (B) (C) (D)	66 (A) (B) (C) (D)	91 (A) (B) (C) (D)	116 (A) (B) (C) (D)	141 (A) (B) (C) (D)
17 (A) (B) (C) (D)	42 (A) (B) (C) (D)	67 (A) (B) (C) (D)	92 (A) (B) (C) (D)	117 (A) (B) (C) (D)	142 (A) (B) (C) (D)
18 (A) (B) (C) (D)	43 (A) (B) (C) (D)	68 (A) (B) (C) (D)	93 (A) (B) (C) (D)	118 (A) (B) (C) (D)	143 (A) (B) (C) (D)
19 (A) (B) (C) (D)	44 (A) (B) (C) (D)	69 (A) (B) (C) (D)	94 (A) (B) (C) (D)	119 (A) (B) (C) (D)	144 (A) (B) (C) (D)
20 (A) (B) (C) (D)	45 (A) (B) (C) (D)	70 (A) (B) (C) (D)	95 (A) (B) (C) (D)	120 (A) (B) (C) (D)	145 (A) (B) (C) (D)
21 (A) (B) (C) (D)	46 (A) (B) (C) (D)	71 (A) (B) (C) (D)	96 (A) (B) (C) (D)	121 (A) (B) (C) (D)	146 (A) (B) (C) (D)
22 (A) (B) (C) (D)	47 (A) (B) (C) (D)	72 (A) (B) (C) (D)	97 (A) (B) (C) (D)	122 (A) (B) (C) (D)	147 (A) (B) (C) (D)
23 (A) (B) (C) (D)	48 (A) (B) (C) (D)	73 (A) (B) (C) (D)	98 (A) (B) (C) (D)	123 (A) (B) (C) (D)	148 (A) (B) (C) (D)
24 (A) (B) (C) (D)	49 (A) (B) (C) (D)	74 (A) (B) (C) (D)	99 (A) (B) (C) (D)	124 (A) (B) (C) (D)	149 (A) (B) (C) (D)
25 (A) (B) (C) (D)	50 (A) (B) (C) (D)	75 (A) (B) (C) (D)	100 (A) (B) (C) (D)	125 (A) (B) (C) (D)	150 (A) (B) (C) (D)

FIREFIGHTER, F. D.

At the present time in most jurisdictions throughout the United States, municipal employees who were once referred to as firemen are now called firefighters. Therefore, whenever you encounter the title fireman while reviewing the following section you should substitute the word firefighter.

MATHEMATICS IN FIREFIGHTING

1. A water tank has a capacity of 6000 gallons. Connected to the tank is a pump capable of supplying water at the rate of 25 gallons per minute which goes into operation automatically when the water in the tank falls to the one-half mark. If we start with a full tank and drain the water from the tank at the rate of 50 gallons a minute, the tank can continue supplying water at the required rate for
(A) 2½ hours (B) 3 hours
(C) 3½ hours (D) 4 hours
(E) 4½ hours.

2. Three firemen are assigned the task of cleaning fire apparatus, which usually takes three men five hours to complete. After they have been working three hours, three additional firemen are assigned to help them. Assuming that they all work at the normal rate, the assignment of the additional men will reduce the time required to complete the task by
(A) 20 minutes (B) 30 minutes
(C) 40 minutes (D) 50 minutes
(E) 60 minutes.

3. Engine A leaves its firehouse at 1:48 p.m. and travels 3 miles to a fire at an average speed of 30 miles per hour. Engine B leaves its firehouse at 1:51 P.M. and travels 6 miles to the same fire at an average speed of 40 miles per hour. From the above facts, we may conclude that Engine A arrives
(A) 3 minutes before Engine B
(B) 6 minutes before Engine B
(C) 3 minutes after Engine B
(D) 6 minutes after Engine B
(E) at the same time as Engine B.

4. Assume that at the beginning of the calendar year an employee was earning $4,800 per year. On July 1st, he received an increase of $240 per year. On November 1st, he was promoted to a position paying $6,000 per year. The total earnings for the year were most nearly
(A) $5100 (B) $4900
(C) $5000 (D) $5400
(E) $5300.

5. A fire engine carries 900 feet of 2½″ hose, 500 feet of 2″ hose and 350 feet of 1½″ hose. Of the total hose carried, the percentage of 1½″ hose is most nearly
(A) 35 (B) 30
(C) 25 (D) 20
(E) 15.

6. An engine company made 96 runs in the month of April which was a decrease of 20% from the number of runs made in March. The number of runs made in March was most nearly
(A) 124 (B) 80
(C) 120 (D) 110
(E) 112.

7. Suppose that a fire truck is undergoing tests. If the distance traversed by the truck, where the distance is expressed in yards, is divided by the time required by the truck to cover that distance, where the time is expressed in minutes, then the quotient is equal to the
(A) maximum rate of speed at which the truck travelled during any one minute
(B) average rate of speed of the truck measured in yards per minute
(C) average distance travelled by the truck, measured in yards
(D) number of miles per hour at which the truck travelled
(E) total distance travelled by the truck, measured in miles.

8. If an engine pumps G gallons of water per minute, then the number of gallons pumped in half an hour may be found by
(A) taking one-half of G
(B) multiplying G by 60 and then dividing the product by two
(C) dividing 60 by twice G
(D) dividing G by 30 and then multiplying the quotient by two
(E) dividing 30 by G.

9. Assume that six men are required to operate a piece of apparatus and that each man is on duty 42 hours per week. Assume further that time lost from duty because of vacations, sick leave and other reasons amounts to 10 per

cent of the company's manpower. Under these conditions, the number of men required to operate this apparatus at full strength around the clock is most nearly
(A) 21 (B) 24
(C) 27 (D) 30.

10. The spring of a spring balance will stretch in proportion to the amount of weight placed on the balance. If a 2-pound weight placed on a certain balance stretches the spring 1/4″, then a stretch in the spring of 1-3/4″ will be caused by a weight of
(A) 10 lbs. (B) 12 lbs.
(C) 14 lbs. (D) 16 lbs.

11. A fireman can inspect X number of buildings in Y number of days. In W number of days he can inspect
(A) $\frac{X}{YW}$ (B) $\frac{XW}{Y}$
(C) $\frac{W}{XY}$ (D) $\frac{Y}{XW}$.

12. Two fire houses are located on the same avenue 10 miles apart. A house located on this avenue somewhere between the two companies catches fire and both companies respond, starting at exactly the same time and reaching the fire at exactly the same time. One engine travelled at an average speed of 20 miles per hour and the other at an average speed of 30 miles per hour. The number of miles travelled by the faster engine was
(A) 5 (B) 6
(C) 7 (D) 8.

13. In the past year, fires occurred in K "Type Z" multiple dwellings. It is known that L percent of the M multiple dwellings are of "Type Z." The fraction of "Type Z" multiple dwellings in which fires occurred during the past year is
(A) K divided by L times M
(B) L times M divided by 100 K
(C) K divided by the quantity 100 times L times M
(D) 100 K divided by the quantity L times M
(E) none of the foregoing.

14. In the local Fire Department there are A firefighters, D lieutenants, E captains, and G chiefs of various ranks. Suppose that, for comparative purposes, promotional opportunities are evaluated as the ratio of the number of promotional positions to the number of positions at the entrance level. In accordance with this method, promotional opportunities in the uniformed force of the Fire Department are evaluated as
(A) G divided by the sum of A plus D plus E
(B) the sum of D plus E plus G divided by the number of firemen
(C) A divided by the sum of D plus E plus G
(D) the sum of A plus D divided by the sum of E plus G
(E) a fraction about which it is known only that the numerator is greater than the denominator.

15. Suppose that the amount of money that the local Fire Department has saved its citizens is estimated at P dollars. If this sum is increased at least 100 per cent in the next year, then the savings must be at least
(A) equivalent to the ratio between P and 100
(B) commensurate with a sum derived by arithmetic manipulation involving P, 100, and a third value not given in the problem
(C) 100 times P dollars
(D) twice P dollars
(E) a sum of money not accurately described in any of the foregoing options.

16. Suppose that a ladder consists of four sections, each R feet in length. When the ladder is extended, adjacent sections overlap for a distance of S feet to strengthen the interlocking. The total overall length of the ladder, when fully opened, is
(A) 4 R feet
(B) 4 R minus 3 S feet
(C) 4 R minus 4 S feet
(D) 4 R minus 6 S feet
(E) none of the foregoing.

17. The velocity of a fire engine which is traveling to a fire is computed by
(A) multiplying distance by time
(B) dividing distance by time
(C) squaring the force with which the earth attracts the engine
(D) means of the moment of inertia
(E) use of the Pythagorean Theorem.

18. Suppose that R persons were rescued from burning buildings by firefighters in March. Suppose also that P persons perished in burning buildings in April. If R is less than S but greater than T, and P is less than both M and N, it may safely be concluded that
(A) the sum of R and T is greater than S
(B) the sum of M and N is greater than P
(C) R is between M and N times as great as P
(D) R exceeds P to an indeterminate degree lying somewhere between S and N
(E) none of the foregoing options is correct.

19. Suppose that the number of fire occurring in a particular type of dwelling decreased C per cent in July, as compared to June, but then increased C per cent in August, as compared to July. Then the number of fires occurring in that type of dwelling during August, as compared to June, is
(A) decreased by the per cent equal to C squared divided by 100
(B) unchanged
(C) increased by the per cent equal to the fraction whose numerator is 100 minus C and denominator is 100
(D) decreased by the per cent equal to the fraction C over 100

20. Suppose that ½ of the firemen in the Fire Department have served for more than 5 years and ⅓ for more than 10 years. Then the fraction of firemen having served between 5 and 10 years is most nearly
(A) ⅔ (B) ⅓
(C) ⅕ (D) ⅙
(E) 1/12.

21. Suppose that a piece of fire apparatus averages 20 runs a month during each of 3 summer months and 30 runs a month during each of 2 winter months. Then the average number of runs made by this apparatus over the 5 month period is most nearly
(A) 22 (B) 24
(C) 25 (D) 26
(E) 28.

22. A basement is being flooded by means of fire hose. At the end of 6 minutes, the basement is one-third flooded. At the same rate of flooding, a basement with a capacity for holding twice as much water could be completely flooded in
(A) 12 minutes (B) 18 minutes
(C) 24 minutes (D) 36 minutes
(E) 64 minutes.

23. Assume that, during a certain period, there are 300 fires and 200 false alarms of fires. If the number of fires were doubled and the number of false alarms were also doubled, the ratio of fires to false alarms would be
(A) four times as large
(B) twice as large
(C) unchanged
(D) one-half as large
(E) one-quarter as large

24. Assume that, during a given period of time, the number of third alarms was 80 per cent of the number of second alarms. Then, of the following, the most accurate statement is that
(A) the number of second and third alarms combined was at least 80 per cent of all first alarms
(B) at least 80 per cent of all alarms higher than first alarm were second alarms
(C) the number of second alarms was 20 per cent of the number of third alarms
(D) there were one-quarter as many third as second alarms
(E) there were 1¼ times as many second as third alarms.

25. Assume that, of the 130 runs made by Fire Department apparatus during a certain period of time, 50 per cent were made by engine companies, 40 per cent by truck companies, and the remainder by miscellaneous companies. Then, of the following, the most accurate statement is that
(A) truck companies made fewer than 50 runs
(B) engine companies made more than 75 runs
(C) truck companies made 10 runs more than engine companies
(D) miscellaneous companies made more than 10 runs
(E) engine companies and truck companies together made fewer than 100 runs.

In Column I are five problems, items 26 to 30. In Column II are twenty numbers. For each problem in Column I select the number from Column II which is the answer to that problem. Place the letter next to the number in the properly numbered space on the answer sheet.

Column I

26. A hose line 1,000 feet long is connected to a hydrant. Hydrant pressure is 96 pounds. What is the the nozzle pressure in pounds? The formula for finding nozzle pressure when the

hydrant pressure is known is as follows:

$$N = \frac{H}{1.1 + KL}, \text{ where}$$

N = nozzle pressure
H = hydrant pressure
K = .105
L = the number of 50 foot lengths of hose in the line.

27. This year, fires in a certain type of occupancy occurred as follows: 3,449 in Manhattan, 2,979 in the Bronx, 3,353 in Brooklyn, 2,784 in Queens, 575 in Richmond.
If you assume 365 days per year, what was the average daily number of fires in this type of occupancy?

28. A ladder consists of five sections, each 30 feet in length. In order to strengthen the interlocking when it is extended, adjacent sections overlap for a distance of 18 inches. When the ladder is extended to its full length, how many feet will be lost due to overlapping?

29. Ten firemen were asked to carry a load weighing 520 pounds. If three more firemen had helped them, how much less would each fireman carry?

30. Assume that a ladder 30 feet in length is placed against a building sc that the bottom of the ladder is 18 feet from the building and the top of the ladder rests against a window sill. How many feet above the ground is the top of the ladder?

Column II

(A)	6	(E)	14	(I)	22	(O)	30	(V)	38
(B)	8	(F)	16	(L)	24	(P)	32	(W)	40
(C)	10	(G)	18	(M)	26	(S)	34	(X)	42
(D)	12	(H)	20	(N)	28	(T)	36	(Z)	44

31. The zero point of a centigrade scale is equivalent to 32 degrees scale on a Fahrenheit scale and 100 degrees of the centigrade is equal to 212 degrees of the Fahrenheit scale. If a centigrade thermometer reads 15 degrees, the equivalent readings in degrees on a Fahrenheit thermometer is
(A) 32 (B) 27
(C) 40 (D) 59
(E) impossible to compute without additional data.

32. If K fires break out during a given month and result in P dollars of loss, the average loss per fire in dollars is
(A) K multiplied by P
(B) P divided by K
(C) K divided by P
(D) 12 times K divided by 2 times P
(E) none of the foregoing.

33. The annual salary of fireman I is R dollars less than that of fireman II. The second fireman earns V dollars annually. The amount of monthly salary by which the second fireman exceeds the first is given by
(A) V minus R divided by 12
(B) R times V divided by 12
(C) V minus R
(D) 12 times V minus R
(E) R divided by 12.

Mathematics

1. B	8. B	15. E	22. D	29. D
2. E	9. C	16. B	23. C	30. L
3. B	10. C	17. B	24. E	31. D
4. A	11. B	18. B	25. D	32. B
5. D	12. B	19. A	26. O	33. E
6. C	13. A	20. D	27. T	
7. B	14. B	21. B	28. A	

Fire Prevention and Inspection

PREVIOUS TEST QUESTIONS FOR PRACTICE

1. A fireman inspecting buildings in a commercial area, came to one whose outside surface appeared to be of natural stone. The owner told the fireman that it was not necessary to inspect his building as it was "fireproof." The fireman, however, completed his inspection of the building. Of the following, the best reason for continuing the inspection is that
   (A) stone buildings catch fire as readily as wooden buildings
   (B) the Fire Department cannot make exceptions in its inspection procedures
   (C) the building may have been built of imitation stone
   (D) interiors and contents of stone buildings can catch fire.

2. It has been suggested that property owners should be charged a fee each time the Fire Department is called to extinguish a fire on their property. Of the following, the best reason for rejecting this proposal is that
   (A) delay in calling the Fire Department may result
   (B) many property owners don't occupy the property they own
   (C) property owners may resent such a charge as they pay real estate taxes
   (D) it may be difficult to determine on whose property a fire started.

3. Steel supporting beams in buildings often are surrounded by a thin layer of concrete to keep the beams from becoming hot and collapsing during a fire. The one of the following statements which best explains how collapse is prevented by this arrangement is that concrete
   (A) becomes stronger as its temperature is increased
   (B) acts as an insulating material
   (C) protects the beam from rust and corrosion
   (D) reacts chemically with steel at high temperatures.

4. "Smoldering cigarette butts are said to cause many fires." "Smoldering" means most nearly
   (A) blazing furiously
   (B) immersed in oil
   (C) burning in a smothered way
   (D) sooty
   (E) carelessly thrown about.

5. Suppose that a neighbor, knowing that you are a fireman, were to ask you whether there is more hazard in the use of kerosene than gasoline at ordinary room temperatures. You should reply that there is more hazard in the use of
   (A) kerosene, because it gives off dangerous quantities of explosive vapors which are lighter than air
   (B) gasoline, because gasoline vapor may flow along the floor and be ignited at a long distance from its point of origin
   (C) kerosene, because its flash point is very low
   (D) gasoline, particularly because when ignited it burns
   (E) kerosene, because its vapors are not easily detected by the sense of smell and yet may be present in dangerous quantities.

6. "Vertical openings such as dumbwaiters, elevators, and chutes are the bane of a fire fighting force." This condition arises mainly because the existence of such openings in a burning building facilitates
   (A) accidental falls
   (B) generation of gases
   (C) spread of the fire
   (D) the perpetration of arson
   (E) collapse of wall supports.

7. Suppose that a factory has stored within it a number of substances. If the owner asked you which of the following is most likely to constitute a fire hazard, you would reply
   (A) sodium chloride
   (B) calcium chloride
   (C) chromium
   (D) silicon dioxide
   (E) absorbent cotton.

8. The suggestion has been made that groups of firemen, without apparatus of any kind, be kept in reserve at a few centrally located points throughout the city. Of the following, the most valid justification for this proposal is that
   (A) when second or third alarms are sent, the need is often for more men rather than more apparatus

(B) the Fire Department is undermanned
(C) the fire districts in a city should be revised periodically to meet population trends
(D) discipline is quite as important as apparatus in extinguishing fires quickly
(E) men are as important as machines in the fighting of fires.

9. At a meeting concerned with fire prevention, it was said: "The fact that the fire loss has been maintained near to its previous levels is encouraging evidence that our increasing efforts over the years in the field of public education in fire protection have not been unavailing." Of the following, the most essential assumption that must be made if this statement is accepted is that
(A) further public education in fire protection is desirable
(B) fire loss has been computed upon the basis of real value, rather than insured value
(C) reference is made here to losses due to fires caused by carelessness, rather than sabotage
(D) there has been in recent years an increase in the potential fire hazard
(E) while the fire loss in recent years has not been reduced greatly, there has been a steady trend in that direction.

10. If you were a firefighter assigned to a waterfront warehouse district, you would generally expect the proportion of daytime fires which developed into multiple alarm blazes to be smaller than the corresponding proportion in fires occurring at night because
(A) people tend to be more careless concerning how they dispose of matches and cigarettes when it is dark
(B) fires have an opportunity to progress farther at night before they are detected and reported
(C) traffic is lighter at night, making it possible to concentrate a larger amount of fire apparatus more quickly
(D) night watchmen in waterfront warehouses are not generally trained in modern methods of fire fighting
(E) warehouses are commonly used to store highly inflammable material.

11. From the viewpoint of fire safety, the chief advantage of a foam rubber mattress compared to a cotton mattress is that the foam rubber mattress
(A) is slower burning
(B) generates less heat when burning
(C) does not smolder
(D) is less subject to water damage.

12. The one of the following methods of storing large piles of coal which is undesirable because it increases the danger of spontaneous heating is
(A) making the pile compact by use of a roller
(B) storing the coal on smooth, solid ground
(C) covering the sides and top of the pile with road tar
(D) mixing coal of various sizes in one pile.

13. "After a fire has been extinguished, every effort should be made to determine how the fire started." Of the following, the chief reason for determining the origin of the fire is to
(A) reduce the amount of damage caused by the fire
(B) determine how the fire should have been fought
(C) eliminate causes of fire in the future
(D) explain delays in fighting the fire
(E) improve salvage operations.

14. "A partially filled gasoline drum is a more dangerous fire hazard than a full one." Of the following, the best justification for this statement is that
(A) a partially filled gasoline drum contains relatively little air
(B) gasoline is difficult to ignite
(C) when a gasoline drum is full the gasoline is more explosive
(D) gasoline vapors are more explosive than gasoline itself
(E) air is not combustible.

15. "Chief officers shall instruct their Aides that when transmitting particulars of a fire, they shall include the fact that foods are involved if the fire involves premises where foodstuffs are sold or stored." Of the following, the best justification for this regulation is that, when foodstuffs are involved in a fire
(A) the fire may reach serious proportions
(B) police protection may be desirable to prevent looting
(C) relatively little fire fighting equipment may be needed
(D) there is a strong likelihood of arson
(E) inspection to detect contamination may be desirable.

16. The one of the following circumstances concerning a fire which indicates most strongly

the possibility of arson is that
(A) there was heavy charring of wood around the point of origin of the fire
(B) three fires apparently broke out simultaneously in different parts of the building
(C) the heat was so intense that glass in the building became molten and fused
(D) when the firemen arrived the smoke was very heavy
(E) the fire apparently started in an oil soaked mop.

17. As a fireman, you may be assigned to inspect buildings for fire hazards. The one of the following most appropriately used for fire-retardant coating of wood is
(A) varnish (B) shellac
(C) wood stain (D) lacquer
(E) white wash.

18. "The taxpayer type of building is usually located in the busiest sections of a city." Of the following, the *least* valid implication of the above statement is that, when a fire in a taxpayer occurs during working hours
(A) heavy traffic may be a handicap to fire fighting operations
(B) precautions may be necessary to avoid panic among employees or shoppers
(C) police lines may be needed to keep back pedestrians
(D) failure to get the fire under control quickly may result in high property damage
(E) the fire may burn for a long period of time before it is detected.

19. Of the following, the basic purpose of Fire Department inspection of private property is to
(A) make sure that Fire Department equipment is properly maintained
(B) secure proper maintenance of all entrance and exit facilities
(C) obtain correction of conditions creating undue fire hazards
(D) make sure that walls are properly plastered and painted in order to prevent the spread of fire
(E) impress upon owners and occupants the importance of fire prevention.

20. The one of the following which is not a *fire resister* is
(A) asbestos (B) tungstate of soda
(C) borax (D) sulphur
(E) phosphate of ammonia.

21. Fire drills should be held frequently in school buildings chiefly in order to
(A) impress upon pupils the necessity for discipline
(B) conform to the rules of the Fire Department and the Board of Education
(C) acquaint pupils with Fire Department regulations
(D) train pupils to leave the building quickly and in an orderly manner
(E) lead pupils to respect law and order.

22. The exit doors in a theatre should swing out in the direction of the street mainly because
(A) panics should be avoided
(B) people should walk, not run, to the nearest exit
(C) the doors may catch fire
(D) exits are then more readily seen
(E) audiences can then get out easier.

23. Suppose that you are a fireman making a routine inspection of a rubber goods factory. During the inspection you discover some minor violation of the Fire Prevention Code. When you call these violations to the attention of the factory owner, he becomes annoyed and tells you that he is the personal friend of high officials in the Fire Department and the city government. Under these circumstances, the best of the following courses for you to follow is to
(A) summon a policeman to arrest the owner for attempting to intimidate a public official performing his duty
(B) make a very thorough inspection and serve summonses for every possible violation of the Fire Prevention Code
(C) ignore the owner's remarks and continue the inspection in your usual manner
(D) try to obtain from the owner the names and positions of his friends.

24. Spontaneous combustion may be the reason for a pile of oily rags catching fire. In general, spontaneous combustion is the direct result of
(A) application of flame
(B) falling sparks
(C) intense sunlight
(D) chemical action
(E) radioactivity.

25. The one of the following statements about electric fuses that is most valid is that they
(A) should never be replaced by coins
(B) may be replaced by coins for a short time if there are no fuses available

(C) may be replaced by coins provided that the electric company is notified
(D) may be replaced by coins provided that care is taken to avoid overloading the circuit
(E) may be replaced only by a licensed electrician.

26. The Fire Department has criticized the management of several hotels for failure to call the Fire Department promptly when fires are discovered. The most probable reason for this delay by the management is that
(A) fire insurance rates are affected by the number of fires reported
(B) most fires are extinguished by the hotel's staff before the Fire Department arrives
(C) hotel guests frequently report fires erroneously
(D) it is feared that hotel guests will be alarmed by the arrival of fire apparatus
(E) many fires smolder for a long time before they are discovered.

27. Heavy blizzards greatly increase the problems and work of the Fire Department. When such a situation occurs, the fire commissioner could reasonably be expected to
(A) order members of the Fire Department to perform extra duty
(B) limit parking on city streets
(C) station firemen at fire alarm boxes to prevent the sending of false alarms
(D) prohibit the use of kerosene heaters.

28. A fireman, taking some clothing to a dry cleaner in his neighborhood, noticed that inflammable cleaning fluid was stored in a way which created a fire hazard. The fireman called this to the attention of the proprietor, explaining the danger involved. This method of handling the situation was
(A) bad; the fireman should not have interferred in a matter which was not his responsibility
(B) good; the proprietor would probably remove the hazard and be more careful in the future
(C) bad; the fireman should have reported the situation to the fire inspector's office without saying anything to the proprietor
(D) good; since the fireman was a customer, he should treat the proprietor more leniently than he would treat other violators
(E) bad; the fireman should have ordered the proprietor to remove the violation immediately and issued a summons.

29. A fireman caught a civilian attempting to re-enter a burning building despite several warnings to stay outside of the fire lines. The civilian insisted frantically that he must save some very valuable documents from the fire. The fireman then called a policeman to remove the civilian. The fireman's action was
(A) wrong; it is bad public relations to order people about
(B) right; the fireman is charged with the responsibility of protecting lives
(C) wrong; the fireman should have explained to the civilian why he should not enter the building
(D) right; civilians must be excluded from the fire zone
(E) wrong; every person has a right to risk his own life as he sees fit.

Items 30 to 32 inclusive, are based on the following paragraph:

"A flameproof fabric is defined as one which, when exposed to small sources of ignition, such as sparks or smoldering cigarettes, does not burn beyond the vicinity of the source of the ignition. Cotton fabrics are the materials commonly used that are considered most hazardous. Other materials, such as acetate rayons and linens, are somewhat less hazardous, and woolens and some natural silk fabrics, even when untreated, are about the equal of the average treated cotton fabric insofar as flame spread and ease of ignition are concerned. The method of application is to immerse the fabric in a flameproofing solution. The container used must be large enough so that all the fabric is thoroughly wet and there are no folds which the solution does not penetrate."

30. According to the above paragraph, a flameproof fabric is one which
(A) is unaffected by heat and smoke
(B) resists the spread of flames when ignited
(C) burns with a cold flame
(D) cannot be ignited by sparks or cigarettes
(E) may smolder but cannot burn.

31. According to the above paragraph, woolen fabrics which have not been flameproofed are as likely to catch fire as
(A) treated silk fabrics
(B) treated acetate rayon fabrics
(C) untreated linen fabrics
(D) untreated synthetic fabrics
(E) treated cotton fabrics.

32. In the method described above, the flameproofing solution is best applied to the fabric by
(A) sponging the fabric

(B) spraying the fabric
(C) dipping the fabric
(D) brushing the fabric
(E) sprinkling the fabric.

33. The real problem in fire prevention and protection in the presence of radioactive isotopes is based on the fact that the radioactivity of an element or compound
(A) is not diminished by change of the compound by fire or explosion
(B) increases directly as the intensity of fire in establishments using them
(C) causes a compound to vaporize, melt or oxidize more readily
(D) is reduced significantly if control by means of suitable sprays or extinguishers is immediately applied
(E) increases with the decrease in the size of the radioactive particle.

34. The most prolific structural cause of fire spread in non-fireproof hotels, other than that of unprotected vertical openings is
(A) inadequate auxiliary appliances and alarm systems
(B) unprotected horizontal openings
(C) poorly protected storage of paints, oils, wastes and volatile mixtures
(D) inadequately protected linen and furniture storage rooms
(E) large dining areas, cocktail lounges, etc. with combustible decorations

35. Of the following metals, the one which is *least* acceptable as a non-sparking metal for tools is:
(A) hardened copper (B) bronze
(C) brass (D) copper alloys.

36. Of the heating defects responsible for hotel fires, the major defect is
(A) defective flues
(B) overheated appliances
(C) defective appliances
(D) inadequate clearance.

37. A television receiver has a greater inherent fire hazard than a conventional radio receiver because
(A) of greater electrical leakage
(B) cabinets are inadequately ventilated
(C) higher voltage is used in the system
(D) they are operated for longer periods of time
(E) the coaxial cable lead-in is covered with a highly flammable coating.

38. In an inspection of a medical laboratory, the fire officer found that ethyl ether was stored in a modern domestic household refrigerator to reduce vaporization possibilities. Of the following, the most likely reason for considering this method of storage inadvisable is that
(A) desired temperature ranges for such storage are not normally available in domestic refrigerators
(B) released gases or vapors may escape without detection in such storage spaces
(C) explosion vents are usually lacking in the refrigerated storage space
(D) exposed electrical parts of the refrigerator are usually located in the storage space
(E) the condensing unit of the refrigerator is easily accessible in the refrigerated storage space.

39. The one of the following combustible materials whose tendency to spontaneous heating is very slight is
(A) lamp black (B) fish meal
(C) cotton seed (D) waste paper
(E) roofing felt.

40. If, upon a reinspection of a plant which has 30 days to comply with a previous order, you find that the order has not been completely obeyed but that some work has taken place, you should
(A) report to proper authorities to obtain legal action
(B) assume the delay is unavoidable and check again in 30 days
(C) inform the person in charge that a 10 day extension will be granted and that legal action will be taken if the order has not been followed
(D) do none of the above.

41. An analysis of loss records in one city showed that one third of the total loss in building fires was in residence buildings and that of the total loss in such buildings, in the year under study, nearly 80% was in multiple dwelling buildings. The one of the following courses of action for the fire department which should be taken immediately on the basis of this report is to
(A) relocate companies
(B) recommend sprinkler protection for multiple dwellings
(C) institute special training in fighting multiple dwelling fires
(D) reduce the protection given to other than multiple dwelling residences
(E) inspect multiple dwellings more thoroughly.

42. "It is good practice to so install heating devices that under conditions of maximum heat (long continued exposure) they will not cause the temperature of exposed woodwork to exceed 160° F." This practice is
(A) correct because of the possibility that wood and other combustible materials after long continued exposure to relatively moderate heat may ignite at temperatures far below their usual ignition temperatures
(B) not correct because no wood in ordinary use will ignite at a temperature of less than 400° F. and consequently the requirement is needlessly severe
(C) correct because oxidation proceeds much more rapidly at higher temperatures
(D) not correct because oxidation proceeds much more slowly at higher temperatures
(E) correct because under prolonged heating the temperature of the air in the room will build up until the ignition point is reached unless the applied temperature is kept sufficiently low.

43. Analysis of the causes of fires is important, as only by knowing the causes of fires is it possible to effectively prevent fire. An analysis of fires in rooms used for spraying flammable paints and finishes has shown that the most important of the following causes is
(A) smoking by employees
(B) defective electrical equipment
(C) spontaneous ignition of paint deposits, rubbish, wiping rags
(D) static electricity resulting from friction
(E) cutting and welding operations.

44. The flammability or combustibility of radioactive materials has little or no direct effect on the fire hazard of a laboratory, primarily because
(A) the unusual structural characteristics of such a laboratory serve to limit possible fire spread from such hazards
(B) water or water spray are effective on most radioactive substances
(C) the quantities of such material in any one laboratory are usually small
(D) laboratory fire prevention and fire fighting facilities usually exceed maximum fire hazards
(E) such materials are inherently of a low order of combustibility.

45. Comparison of the burning qualities of foam rubber and cotton mattresses shows that, generally,
(A) a cotton mattress burns faster but cooler
(B) a foam rubber mattress burns slower but hotter
(C) a foam rubber mattress burns faster and hotter
(D) a cotton mattress burns faster and hotter
(E) the potential fire hazard of a foam rubber mattress is higher.

46. During Christmas and other holiday shopping seasons, it is required that frequent inspections shall be made of department stores, at irregular intervals. Of the following, the most important reason for this inspection procedure is to
(A) prevent unnecessary interference with store operations
(B) check characteristic holiday operations
(C) permit more frequent and thorough coverage of the stores in question
(D) avoid delay in urgent fire operations
(E) permit more flexible scheduling of inspections.

47. The fire load computation of a building indicates, for the most part, the
(A) risk of a fire breaking out
(B) rate at which a fire is likely to grow
(C) combustibility of the various parts of the building rather than its contents
(D) amount of combustibles and the method of protection
(E) maximum fire stress to which the building might be subjected.

48. "It is far more important that escape routes from multi-story buildings should be protected against smoke and hot gases than from direct flame or heat." This statement is
(A) not correct; resistive construction is likely to be smoke resistive as well
(B) correct; adequate means for ventilation is essential to prevent cutting off escape routes
(C) not correct; unless corridors and escape stairs are constructed of fire resistive materials, progress of fire cannot be blocked
(D) correct; unless properly protected from smoke and hot gases, escape stairs would be unusable by occupants
(E) not correct; unless properly protected against direct flame or heat, escape routes cannot resist smoke or hot gases.

49. The minimum width of exit usually required for a single file of persons is most nearly
   (A) 15-17 inches (B) 18-20 inches
   (C) 21-23 inches (D) 24-27 inches
   (E) 27-30 inches.

50. The most pronounced method of reducing the fire and life hazards in public buildings is by
   (A) ample exits of any type
   (B) ample stairways inside the buildings
   (C) the use of exit signs and panic locks
   (D) fire-resisting construction suitable for the occupancy.

51. From all viewpoints, the most hazardous materials that could be stored, that would be a life as well as a fire hazard would be
   (A) second hand niter bags
   (B) used motor vehicles
   (C) loose or baled vegetable fibers
   (D) pyroxylin and pyroxylin plastic products.

52. The chief fire hazard of welding and cutting operations is
   (A) flames of the torch igniting nearby material
   (B) broken hose line
   (C) flying sparks
   (D) backfiring of torch.

53. Of the following items associated with motion picture theatres, the primary hazard is
   (A) misuse of electricity
   (B) heating defects
   (C) smoking and matches
   (D) projection booth fires.

54. The most hazardous method of fumigation is by the use of
   (A) heat (125°F.)
   (B) carbon tetrachloride mixture with flammable fumigant
   (C) carbon dioxide mixture with flammable fumigant
   (D) carbon bisulphide.

55. The combined use of inspections, periodic reports of activities, follow-up procedures, special reports from subordinates and a rating system comprise a system of
   (A) coordination (B) command
   (C) control (D) representation
   (E) on-the-job training.

56. What should be provided in air conditioning ducts to prevent the spread of fire and smoke through a property?
   (A) automatic dampers
   (B) intake screens
   (C) steel wool air filters
   (D) heat actuated devices.

57. The outstanding fire hazard of boarding and rooming houses is
   (A) misuse of electricity
   (B) smoking and matches
   (C) heating defects
   (D) incendiary.

58. The simplest and most feasible method of avoiding the overheating of woodwork near any high temperature heating appliance is by
   (A) filling the intervening space with insulating material
   (B) covering the woodwork with sheet metal
   (C) providing an air space between the woodwork and the appliance
   (D) covering the woodwork with asbestos sheets.

59. The principal source of fire hazard in connection with heating equipment in mercantile buildings comes from
   (A) defective wiring
   (B) insufficient clearance from combustible materials
   (C) the storage and handling of fuel
   (D) defective motors.

60. "More ink has been spilled on the item of smoking as a cause of fire than on any other, but the total result has been negligible." This situation is best accounted for by the fact that
   (A) smoking is generally an automatic act performed unthinkingly
   (B) truly effective facilities for elimination of smoking hazards are exceedingly cumbersome or expensive
   (C) most people regard smoking as a personal prerogative and resent control measures
   (D) smoking is practiced by many individuals with defective intelligence and social attitudes.

61. There are two basic factors in assessing building construction from the fire prevention standpoint. One is the combustibility of materials. The other is
   (A) extinguishment facilities
   (B) excess structural strength
   (C) ventilation control
   (D) limitation of fire spread
   (E) means of access and egress.

62. The *least* accurate statement concerning the protection of openings in walls and partitions is
   (A) protection of wall openings may prevent

either the horizontal or the vertical spread of fire
(B) the general features of a building have no bearing on the extent to which such protection is necessary
(C) the protection secured by fire doors and fire windows can not be better than fire-retardant value of the walls
(D) good solid walls are preferable to those with fire doors in restricting the spread of fire.

63. The materials of which a building are constructed and the opportunities for the spread of fire are important, but the greatest single hazard is usually that of
(A) occupancy
(B) location
(C) fire protective measures
(D) construction.

64. A warehouse with a leaky roof contains a large amount of building material. The one of the following materials which is most likely to set fire to the warehouse is
(A) gasoline (B) crude oil
(C) lime (D) kerosene.

65. For the most effective results in conducting a Fire Prevention Week campaign, it would be desirable to emphasize fire prevention
(A) in its broader community aspects
(B) as a means of lowering insurance rates
(C) as it applies to the individuals' own homes
(D) as a means of lowering the operating costs of the Fire Department
(E) as a means of assuring the uninterrupted operation of business.

66. Floor or wall openings sometimes prevent the banking up of heated air. This condition, with respect to sprinklers, is considered
(A) advantageous (B) unimportant
(C) detrimental (D) good ventilation.

67. The best all-round fireproofing material, due to its high resistance to heat, its lightness, its great strength, its adaptability to any shape and which is also very easily repaired when damaged by a severe fire is
(A) brick (B) hollow clay tile
(C) gypsum (D) concrete.

68. An unusually large number of fires of "unknown cause" is characteristic of the fires involving
(A) restaurants (B) warehouses
(C) mercantile stores (D) hospitals.

69. It is obvious that where a division wall is not continued through the roof, and where the roof is combustible on both sides of the wall, fire is almost certain to spread beyond the wall if the fire is of any duration. According to the foregoing statement there is need of
(A) parapets
(B) more resistive division walls
(C) fire-stopping of roof spaces
(D) fire-stopping of division walls.

70. "The hazard of flammable gases is generally ________ to that of flammable liquids." The above missing word is
(A) opposite (B) dissimilar
(C) similar (D) identical.

71. The most important factor that would materially decrease large-loss supermarket fires is
(A) education of the general public
(B) separating the utility area from the rest of the building bv a fire resistive wall
(C) maintenance of an adequate supply of fire extinguishers
(D) keeping all aisles clear of merchandise and storage
(E) fireproof disposal containers for matches, cigarettes, etc., in the sales area.

72. The primary difference between the large number of small fires that produces a small percentage of total losses, and the smaller number of large fires that accounts for 95% of the total loss in the U. S., is usually
(A) nature of the material involved in the fire
(B) type of structure involved
(C) early discovery
(D) availability of personnel to fight the fire
(E) availability of adequate apparatus.

73. From the fire prevention standpoint, air conditioning, and air blower systems are of concern mainly because they
(A) provide a means for the spread of fire through the building served
(B) severely limit adequate ventilation in case of fire
(C) intensify fires from other sources by providing abnormally large amounts of air
(D) characteristically accumulate hazardous quantities of dust and lints which are subject to spontaneous ignition.

74. At what interval of time should rubbish and waste materials be removed from piers, docks and wharves?
   (A) at least daily
   (B) once per week
   (C) as often as needed to prevent dangerous condition
   (D) as fast as accumulated.

75. Of the various rooms found in the average school building, which two places deserve more consideration from a point of view of preventing personal injuries that may result from panic?
   (A) the auditorium and the boiler room
   (B) the classroom on the highest floor and the room in the lowest (basement) part of the building
   (C) the auditorium and the cafeteria
   (D) the classroom nearest the auditorium exit and the auditorium itself.

76. The underlying reason behind routine periodic and frequent fire prevention inspection is
   (A) that occupants, hazards, and code compliances may vary considerably in given buildings over a short period
   (B) the need for favorable public opinion
   (C) that a large city usually has many new buildings being constructed
   (D) that most individuals continually and consciously try to evade the fire regulations.

77. The first objective of all fire prevention is
   (A) safeguarding life against fire
   (B) reducing insurance rates
   (C) preventing property damage
   (D) confining fire to a limited area.

78. Which one of the following is the cause of the greatest number of fires?
   (A) electrical wiring
   (B) spontaneous ignition
   (C) sparks on roofs
   (D) smoking and matches.

79. The type of occupancy in which the largest number of fires occurs is
   (A) restaurants and other mercantile establishments
   (B) hospitals, theatres, and other public buildings
   (C) dwellings, including apartments and hotels
   (D) bakeries, cleaning establishments, and other manufacturing plants.

80. Which one of the following factors generally should be given the greatest right in estimating the fire risk in a general or mixed public warehouse?
   (A) availability of water hydrants to the warehouse
   (B) location of warehouse with respect to other buildings in the area
   (C) intensity and direction of prevailing winds in the area
   (D) kind of merchandise stored in warehouse.

81. Of the following, the one which is perhaps the most important year round element in fire prevention in residences is
   (A) proper and regular disposal of combustible waste
   (B) care in the operation of heating systems
   (C) periodic inspections by members of the fire department
   (D) radio announcements calling attention to fire hazards in the home.

82. The fire prevention and fire protection problem resolves itself into three phases, each of which must receive attention. The possibility of human or mechanical failure makes it unsafe to place sole reliance on any one method. If two of these phases are preventing the outbreak of fire and preventing the serious spread of fire, then the third phase would be providing for
   (A) extensive research in the cause and prevention of nonincendiary fires
   (B) the specialized training of fire department personnel at all levels
   (C) the prompt detecting and extinguishing of fire
   (D) ample modern fire-fighting equipment.

83. The largest cause of apartment and tenement house fires is
   (A) smoking and matches
   (B) electrical
   (C) gas stoves and explosions
   (D) heating equipment.

84. Fire loss statistics show that 90% of the losses occur at
   (A) 10% of the fires
   (B) 15% of the fires
   (C) 20% of the fires
   (D) 30% of the fires

85. To best analyze the fire prevention and protection problem in a certain section of the city, the most basic thing that it is necessary to know is

(A) the number of fire companies in the area
(B) the structural and occupancy data of the area
(C) the number of people living in the area
(D) the available water supply for the area.

86. Of the following, the greatest fire hazard in furniture and cabinet shops is
(A) spontaneous ignition
(B) heating systems in buildings
(C) exposure
(D) misuse of electricity.

87. The chief objective of fire inspection when performed by fire fighters in the fire department is to
(A) discover and remedy petty violations of fire prevention ordinances and bad prevention practices
(B) keep the public informed and aware of fire hazards
(C) maintain a direct relationship between the fireman and the public and to keep the public aware that the fire department is on its toes
(D) make possible more efficient and intelligent fire fighting.

88. Which of the following types of occupancy normally offers the least life hazard during an early morning fire:
(A) hotel (B) theater
(C) dwelling (D) apartment house.

89. Because fire fighting requires positive and dramatic action, it has a greater appeal for people than preventive measures. To overcome this attitude a fire prevention program should probably emphasize most
(A) fire prevention saves lives; fire protection saves property
(B) fire prevention is the first law of good fire protection
(C) modern fire fighting equipment and efficient personnel are the two primary aims of the fire department
(D) prevention of loss of life and minimum property damage are the two primary aims of the fire department.

90. Tests made by the Underwriters Laboratories indicate that hazards involved in fighting fires where television receiving sets are involved, as compared with fighting fires where ordinary electrical household appliances are involved, are
(A) greater, because of the high vacuum in the cathode tube
(B) greater, because more shorting occurs in television sets
(C) about the same
(D) less, because of the low current requirement of television sets.

91. In air conditioning installations the feature of greatest importance from a fire hazard standpoint is
(A) the possibility of ignition of the air filters or other equivalent parts of the air conditioning system
(B) the difficulty in locating and reducing the seat due to the inaccessibility of the air ducts
(C) the difficulty of reaching the fan switch and other controls under fire conditions
(D) the possibility of fire and smoke being transmitted by smoke from one section of the building to another.

92. The public's attitude toward fire prevention which is most difficult to overcome is
(A) maliciousness (B) indifference
(C) laziness (D) obstinacy.

93. In a large city the number of fire prevention inspections in each of the various sections of the city should usually be proportional to the
(A) number of inhabitants of the section
(B) size of the section
(C) fire hazards in the section
(D) assessed value of property in the section.

94. In the manufacture of goods from plastics and pyroxylin materials, fires are most likely to be caused by
(A) static sparks during winding or unwinding
(B) sparks from machine tools
(C) careless use of flammable liquids during processing
(D) explosion of flammable vapors during processing.

95. The manufacture of which of the following products is the most hazardous from a fire prevention viewpoint:
(A) artificial leather (B) tobacco
(C) chalk and crayon (D) soap.

96. Inspection agencies maintained by fire insurance companies cannot provide fully adequate inspections because
(A) the company does not allow them to indicate the fire hazard
(B) they never inspect property unless it has been reported as being a fire hazard

(C) they do not have any police power
(D) the main objective of these inspections is to reduce the number of industrial accidents.

97. Of the known causes of fires in clothing factories, the most common cause is
(A) smoking and matches
(B) rubbish and sweepings
(C) electric and gas irons
(D) cleaning fluid explosions.

98. According to the National Fire Protection Association, 60% of fires break out at night. If this is correct, which one of the following would most likely be of the greatest protection to the individual family:
(A) telephone extensions in every bedroom
(B) self-contained unit alarm devices installed in the basement of the home
(C) an alert residential police patrol in cars equipped with radio
(D) automatic fire detectors installed on ceilings and adjacent to any potential fire hazard in the home.

99. The chief advantage of having fire fighters make concurrent fire-prevention and fire fighting inspections is that
(A) the station fire fighters are usually better inspectors than the inspectors from the fire prevention bureau
(B) fire fighters are thus kept busy all the time
(C) one visit to a building may be made to serve two purposes thus saving time and labor
(D) more firemen are on duty in the daytime than at night.

100. When the fire department is conducting concurrent fire fighting and fire prevention inspections from the station houses, the men and officers should be instructed in and told to use all of the following techniques, except one. Which one of the following is the exception:
(A) to arrange a routine for inspection rather than a haphazard approach
(B) to establish friendly relations with the owners and occupants thus incidentally promoting good-will for the fire department
(C) to instruct various employees in mercantile establishments in a given building exactly what corrections should be made on given faulty conditions
(D) to immediately notify superior officers or the fire prevention bureau of unusually hazardous conditions or conditions flagrantly violating the laws and regulations.

Key Answers: Prevention and Inspection

1. D	22. E	43. C	64. C	85. B
2. A	23. C	44. C	65. C	86. A
3. B	24. D	45. C	66. C	87. D
4. C	25. A	46. B	67. B	88. B
5. B	26. D	47. E	68. B	89. B
6. C	27. A	48. D	69. A	90. C
7. D	28. B	49. C	70. C	91. D
8. A	29. B	50. D	71. B	92. B
9. A	30. B	51. D	72. C	93. C
10. B	31. E	52. C	73. A	94. B
11. C	32. C	53. D	74. C	95. A
12. D	33. A	54. D	75. C	96. C
13. C	34. B	55. C	76. A	97. A
14. D	35. C	56. A	77. A	98. D
15. E	36. A	57. C	78. D	99. C
16. B	37. C	58. C	79. C	100. C
17. E	38. D	59. B	80. D	
18. E	39. A	60. A	81. A	
19. C	40. A	61. D	82. C	
20. D	41. E	62. B	83. A	
21. D	42. A	63. A	84. A	

RESPONSIBILITIES AND DUTIES

1. A large fire occurs which you, as a fireman, are helping to extinguish. An emergency arises and you believe that a certain action should be taken. Your superior officer directs you to do something else which you consider undesirable. You should
   (A) take the initiative and follow what you originally thought to be the superior line of action
   (B) think the matter over for a few minutes and weigh the virtues of the two lines of action
   (C) waste no time but refer the problem immediately to another superior officer
   (D) attempt to convince the superior officer that your plan has greater merit than his
   (E) obey orders despite the fact that you disagree.

2. While on duty at a fire, a probationary fireman receives an order from his lieutenant which appears to conflict with the principles of firefighting taught at the fire school. Of the following, the best course of action for the fireman to take is to follow the order and at a convenient time, after the fire, to
   (A) discuss the apparent inconsistency with his lieutenant
   (B) discuss the apparent inconsistency with another officer
   (C) mention this apparent inconsistency in an informal discussion group
   (D) ask a more experienced fireman about the apparent inconsistency.

3. While fighting a fire in an apartment when the occupants are not at home, a fireman finds a sum of money in a closet. Under these circumstances, the fireman should turn over the money to
   (A) a responsible neighbor
   (B) the desk sergeant of the nearest police station
   (C) the superintendent of the apartment house
   (D) his superior officer.

4. Suppose that your company is extinguishing a very small fire in a parked automobile. Your commanding officer directs you to perform some act which, as far as you can see, is not going to help in any way to put out the fire. Of the following, the best reason for obeying his order instantly and without question is that
   (A) the Fire Department is a civil organization
   (B) your officer, after all, has been in the service for a much longer period than you
   (C) without discipline, the efficiency of your company would be greatly reduced
   (D) the first duty of the commanding officer is to command
   (E) since the fire is a small one, you can argue the point with your officer after it is all over.

5. Suppose that you are an off-duty fireman driving your car in the mid-town area. As you cross an intersection you hear sirens and, looking back, see fire apparatus approaching. In this situation, the best action for you to take is to
   (A) attempt to clear a path for the fire apparatus by driving rapidly and sounding your horn
   (B) drive to the next intersection and direct traffic until the apparatus has passed
   (C) permit the apparatus to pass, then follow it closely to the fire, sounding your horn as you drive
   (D) pull to the curb, permit the apparatus to pass, then continue on your way.

6. Assume that you are a fireman on your way home after completing your tour of duty. Just as you are about to enter the subway, a man runs up to you and reports a fire in a house located five blocks away. You recognize the man as a simple-minded but harmless person who frequently loiters around fire houses and at fires. Of the following, the best action for you to take is to
   (A) run to the house to see if there really is a fire
   (B) call in an alarm from a nearby telephone
   (C) ignore the report because of the man's mental condition
   (D) call the police and ask that a radio patrol car investigate the report.

7. Assume that you are a fireman. While you are walking along a quiet residential neighborhood about 3 A.M. on a Sunday morning, another pedestrian calls your attention to smoke coming from several windows on the top floor of a three-story apartment house. Of the following, the best action for you to take immediately is to

(A) race through the house, wake the tenants, and lead them to safety
(B) run to the nearest fire alarm box, turn in an alarm, and then run back into the house to arouse the tenants
(C) run to the nearest fire alarm box, turn in an alarm, and stay there to direct the fire apparatus when it arrives
(D) direct the other man to the nearest alarm box with directions to stay there after sending the alarm, while you go to the apartment from which the smoke is issuing
(E) instruct the other man to help you arouse all the tenants in the building before the fire makes too much headway.

8. Assume that you are a fireman, off duty and in uniform in the basement of a department store. A large crowd is present. There are two stairways, 50 feet apart. A woman whom you cannot see screams "Fire!" You should first
(A) rush outside and sound the alarm at a fire alarm box
(B) find the woman who shouted "Fire!" ascertain where the fire has occurred and proceed to extinguish the fire
(C) jump on top of a nearby counter, order everyone to be quiet and not to move; find out who screamed "Fire!" and reprimand that person publicly
(D) jump on top of a counter, obtain the attention of the crowd, direct the crowd to walk to the nearest stairway, and announce that there is no immediate danger
(E) communicate with the store manager, direct him to sound a fire alarm, find the woman who shouted "Fire!," obtain help in handling the crowd.

Responsibilities and Duties

Correct Answers

1. E	3. D	5. D	7. D
2. A	4. C	6. B	8. D

PUBLIC RELATIONS

1. The most important reason for having members of the Fire Department wear uniforms is to
(A) indicate the semi-military nature of the Fire Department
(B) build morale and esprit de corps of members
(C) identify members on duty to the public and other members
(D) provide clothing suitable for the work performed.

2. A number of prominent municipal officials are extremely interested in the operation of the local Fire Department. These officials frequently arrive at the scene of a fire almost as quickly as the firefighters. It is desirable that the Fire Department welcome the interest of these municipal officials mainly because of the probability that
(A) the officials will acquire considerable technical information
(B) the organization of the Fire Department will be modified
(C) fire administration will remain unaffected
(D) the technical details of fire fighting will be widely popularized
(E) the work and problems of the Fire Department will receive due recognition.

3. The owner of a building at which you helped put out a fire complains bitterly to you that the firemen broke a number of cellar windows even before setting out to extinguish the fire in his cellar. Of the following, the best action for you to take is to
(A) question the validity of the data as described by the owner
(B) request him to put his statement in writing
(C) explain the reason for breaking the windows
(D) suggest that he have the cellar windows made of unbreakable glass
(E) remind him that the firemen had accomplished their purpose; they had extinguished the fire.

4. An elderly man approaches you, as officer on duty, to complain that the noise of apparatus responding to alarms wakes him in the middle of the night. Of the following, the best way for you to handle this situation is to
(A) quickly change the subject since the man obviously is a "crank"
(B) ask the man for specific instances of apparatus making noise at night
(C) explain the need for speed in response of apparatus

(D) promise to avoid making unnecessary noise at night.

5. A fireman who was assigned to deliver a talk before a civic organization first made a detailed outline of his talk. Then, with the help of two members of his company, he prepared several demonstrations and charts. The fireman's procedure was
   (A) good; chiefly because careful preparation increases the effectiveness of a talk
   (B) poor; chiefly because a talk should be guided by the questions and comments of the audience and not follow a pre-determined outline
   (C) good; chiefly because the audience will be impressed by the care that the lieutenant has taken in preparing the talk
   (D) poor; chiefly because the talk will appear rehearsed rather than spontaneous and natural.

6. Suppose that you were assigned in charge of a new headquarters bureau that will have extensive correspondence with the public and very frequent mail contact with division and battalion commands. It is decided at the beginning that all communications from the headquarters office are to go out over your signature. Of the following, the most likely result of this procedure is that
   (A) the administrative head of the bureau will spend too much time preparing correspondence
   (B) execution of bureau policy will be unduly delayed
   (C) subordinate officers will tend to avoid responsibility for decisions based on bureau policy
   (D) uniformity of bureau policy as expressed in such communications will tend to be established
   (E) ability of subordinates to write letters and handle bureau problems will be hampered rather than furthered.

7. Suppose that you have been asked to answer a letter from a local board of trade requesting certain information. You find that you cannot grant this request. Of the following ways of beginning your answering letter, the best way is to begin by
   (A) quoting the laws or regulations which forbid the release of this information
   (B) stating that you are sorry that the request cannot be granted
   (C) explaining in detail the reasons for your decision
   (D) commending the organization for its service to the community
   (E) discussing the difficult problems involved in releasing information.

8. The relationship between the Fire Department and the press is like a "two-way street" because the press is not only a medium through which the Department releases information to the public, but the press also
   (A) is interested in the promotion of the Department's program
   (B) can teach the Department good public relations
   (C) makes the Department aware of public opinion
   (D) provides the basis for community cooperation with the Department
   (E) releases such information from another point of view.

9. Assume that you have been asked to prepare an answer to a request from a citizen of the City addressed to the Chief of Department. If this request cannot be granted, it is most desirable that the answering letter begin by
   (A) telling the person that you were glad to receive his request
   (B) listing the laws which make granting the request impossible
   (C) saying that the request cannot be granted
   (D) discussing the problem presented and showing why the Fire Department cannot be expected to grant the request
   (E) indicating how to go about getting the request by another department.

10. Assume that you have been requested by the Chief of Depuartment to prepare for public distribution a statement dealing with a controversial matter. For you to present the Department's point of view in a terse statement making no reference to any other matter is, in general,
   (A) undesirable; you should show all the statistical data you used; how you obtained the data; and how you arrived at the conclusions presented
   (B) desirable; people will not read long statements
   (C) undesirable; the statement should be developed from ideas and facts familiar to most readers
   (D) desirable; the department's viewpoint should be made known in all controversial matters
   (E) undesirable; the department should not

make statements about controversial matters.

11. The most important function of a good public relations job of a fire department should be in
(A) training personnel
(B) developing an understanding of fire dangers on the part of the public
(C) enacting new fire laws
(D) recruiting technically trained firemen.

12. The primary purpose of a public relations program of an administrative organization should be to develop mutual understanding between
(A) the public and those who benefit by organized service
(B) the public and the organization
(C) the organization and its affiliate organizations
(D) the personnel of the organization and the management.

13. You arrive in your office at 11 A.M., having been on an inspection tour since 8 o'clock. A man has been waiting in your office for two hours. He is abusive because of his long wait, and accuses you of sleeping off a hangover at the taxpayers' expense. You should
(A) say that you have been working all morning, and let him sit in the outer office a little longer until he cools off
(B) tell him you are too busy to see him and make an appointment for later in the day
(C) ignore his comments, courteously find out what his business is and take care of him in a perfunctory manner
(D) explain briefly that your duties sometimes take you out of your office and that an appointment would have prevented inconvenience.

14. During the routine inspection of a building you are told very unfavorable personal comments concerning several top officials of the Department by a citizen. Of the following, it would usually be best for you to
(A) try to change the subject as soon as possible
(B) attempt to convince him that he is in error
(C) advise the citizen that your opinion might be like his, but that you can't discuss it
(D) tell the citizen that it would be more proper for him to put his comments in writing
(E) review the improvements introduced by the present administration.

15. "The fire officer should, as far as operations permit, answer any reasonable questions asked of him by the occupant of the premises involved in a fire." However, of the following, the subject about which he should be most guarded in his comments is the
(A) time of the arrival of companies
(B) methods used to control and extinguish the fire
(C) specific cause of the fire
(D) probable extent of the damage
(E) names of the officer or officers directing the fire forces.

16. "If you want the friendship and cooperation of a man, do not do something for him, but get him to do something for you. His ego has been flattered by your request and, if you follow his advice, he has a vested interest in your project." To what extent is this principle of individual relations applicable to a problem of group relations, such as the development of favorable community attitudes toward the fire prevention activities of the Fire Department:
(A) entirely; maximum public participation should be encouraged
(B) to some extent; the participation of certain groups, in certain functions, is desirable
(C) to a very minor extent; the efforts of community groups should be confined to suggestions, advice, and recommendations
(D) to no degree; this principle is incompatible with the concept of democratic law enforcement.

17. The public is most likely to judge personnel largely on the basis of their
(A) experience
(B) training and education
(C) civic mindedness
(D) manner and appearance while on duty
(E) ability to express themselves with simplicity and clarity.

18. When employees consistently engage in poor public relations practices, of the following the cause is most often
(A) disobedience of orders
(B) lack of emotional control
(C) bullheadedness
(D) poor supervision
(E) outside interference.

19. A citizen's support of a Fire Dept. program can best be enlisted by
(A) minimizing the cost to the community
(B) telling him how backward the community

is in its practices and why such a situation exists
(C) telling him that it is his civic duty to do all that he can to support the program
(D) informing him how the program will benefit him and his family
(E) telling him how another community has benefited from the use of the program.

20. The term "public relations" has been defined as the aggregate of every effort made to create and maintain good will and to prevent the growth of ill will. This concept assumes particular importance with regard to public agencies because
(A) public relations become satisfactory in inverse ratio to the number of personnel employed in the public agency
(B) legislators may react unfavorably to the public agency no matter how its public relations policy is developed
(C) they are much more dependent upon public good will than are commercial organizations
(D) they are tax-supported, and depend on the active and intelligent support of an informed public.

21. Of the following, the proper attitude for an administrative officer to adopt toward complaints from the public is that he should
(A) not only accept complaints but should establish a regular procedure whereby they may be handled
(B) avoid encouraging correspondence with the public on the subject of complaints
(C) remember that it is his duty to get a job done, not to act as a public relations officer
(D) recognize that complaints are rarely the basis for significant administrative action.

22. In a governmental agency, the basic objective of being directly concerned with public relations should be to
(A) promote the most efficient administration of the agency
(B) reduce annual reports which will be acceptable to the public
(C) increase the size of the agency
(D) broaden the "scope and activities" of the agency.

23. Which of the following has the least effect on a city department's public relations:
(A) fluctuations in employment
(B) amount of budget
(C) organization plan of department
(D) size of department.

24. The aspect of public relations which has been most neglected by government agencies, but which has been stressed recently in business and industry, is
(A) use of radio
(B) public speaking
(C) advertising
(D) training employees in personal contacts with the public.

25. The main advantage for good, sound public relations is to
(A) build up a good feeling and understanding between the department and the public
(B) gain public support for wage increases and better working conditions
(C) increase public interest in building projects
(D) attain a friendly and sympathetic press.

26. In dealing with the public it is helpful to know that generally most people are more willing to do that for which they
(A) are not responsible
(B) understand the reason
(C) will be given a little assistance
(D) must learn a new skill

27. A fire department may expect the most severe public criticism
(A) when asking for an increase in the budget
(B) when attempting to float a bond issue for new equipment
(C) after a fire which caused a large loss in property
(D) just after a budget increase has been granted.

Key Answers: Public Relations

1. C	7. B	13. D	19. D	25. A
2. E	8. C	14. A	20. D	26. B
3. C	9. C	15. C	21. A	27. C
4. C	10. C	16. B	22. A	
5. A	11. B	17. D	23. C	
6. D	12. B	18. D	24. D	

FIRE FIGHTING FOR FIREMEN

ALARM SIGNALS

1. A study shows that false alarms occur mostly between noon and 1 P.M., and between 3 and 10 P.M. The most likely explanation of these results is many false alarms are sent by
(A) school children (B) drunks
(C) mental defectives (D) arsonists
(E) accident victims.

2. While visiting the lounge of a hotel, a fireman discovers a fire whtich apparently has been burning for some time and is rapidly spreading. Of the following, the first action for him to take is to
(A) find the nearest fire extinguisher and attempt to put out the fire
(B) notify the desk clerk of the fire
(C) send an alarm from the nearest street alarm box
(D) run throughout the hotel and warn all occupants to evacuate the building.

3. When a fire occurs in Merritt, the alarm transmitted over the fire alarm telegraph system sounds in every fire house in Merritt. Of the following, the best justification for this practice is that
(A) some companies are more efficient than others
(B) to listen to a large number of alarms keeps the firemen alert
(C) for maximally effective disposition of fire apparatus, the staff at the central headquarters should be fully aware of the entire situation at any moment
(D) while certain companies are responding to one alarm, other companies must be prepared to respond to new alarms in the same or nearby districts
(E) firemen have an added incentive to work efficiently when they know that all the other firemen are aware that they are at work.

4. The method of reducing the number of false alarms by education alone is most effective for which one of the following types of individuals who pull false alarms:
(A) mental defectives
(B) inebriates
(C) children of school age
(D) former firemen.

5. Excluding direct enforcement measures, the most promising approach to relief of the municipal false alarm problem seems to be along the lines of
(A) better location and distribution of alarm boxes
(B) replacement of telegraph type systems with telephone systems
(C) improved electrical circuitry, including verification
(D) decentralization of fire alarm headquarters
(E) public relations and education.

6. Of the following, the *least* accurate statement with respect to simultaneous calls is that such calls
(A) apply only to land engine company assignments
(B) always include the preliminary signal 99
(C) apply only to boroughs which operate as a single telegraph unit
(D) always include two box numbers
(E) always apply to first alarm assignments.

7. Before responding with apparatus to a verbal alarm of fire, the officer of a land engine company should send to the central station from the morse key on the alarm circuit in company headquarters
(A) a special call for the full first alarm complement to the street box nearest the location of the fire
(B) a special call for the remainder of the first alarm complement to the street box nearest the location of the fire
(C) a special call for the battalion chief on duty to the street box nearest the location of the fire
(D) a special call for his own company to the street box nearest the location of the fire
(E) an out-of-service signal for his own company.

8. The number of fire companies in a city is chiefly determined by all of the following conditions, except
(A) the number of hose streams likely to be required to handle such fires as may be expected
(B) the manpower and capacity of fire de-

partment pumping apparatus and of ladder service
(C) type of fire alarm system in use
(D) the accessibility of various parts of the city to fire companies.

9. "The fire alarm box is an important element of the City's fire protection system." The one of the following factors which is of *least* value in helping a citizen to send a fire alarm quickly by means of a fire alarm box is that
(A) the mechanism of a fire alarm box is simple to operate
(B) fire alarm boxes are placed at very frequent intervals throughout the City
(C) there are several different types of fire alarm boxes in use
(D) specific directions for sending an alarm appear on each fire alarm box
(E) fire alarm boxes are painted a distinctive color.

10. "One or more public ambulances may be called to a street box by sending on the Morse key therein the following signals in the order given: the preliminary signal 777, the number of the street box, and the number to indicate how many ambulances are wanted." In accordance with these instructions, the proper signal for calling four public ambulances to Box 423 is
(A) 4—777—423 (B) 7777—423
(C) 423—777—4 (D) 777—423—4
(E) 7—423—4.

11. "No fireman is expected to use a measuring stick to determine the exact amount of hose needed to stretch a line from the entrance of the burning building to the seat of the fire. He should be guided generally by the following rule—one length of hose for every story." This rule assumes most directly some degree of uniformity in
(A) fire hazards (B) building construction
(C) causes of fires (D) window areas
(E) street widths.

12. "Installation of a modern fire alarm system will mean smaller fires." Of the following, the best justification for this statement is that
(A) if summoned quickly, firemen can control a fire before it has a chance to spread
(B) if the alarm system is modern, firemen can be given a complete picture of a fire even before they respond
(C) some fires, such as fires resulting from explosions, assume large proportions in a few seconds
(D) most industrial establishments depend on more than one method of transmitting fire alarms
(E) an efficient fire alarm system would discourage arsonists from setting fires intentionally.

13. The one of the following statements about the first alarm response assignment of the second section of a double company that is correct is that when the first section responds to
(A) either a street alarm box or special building alarm, and shortly thereafter an alarm is received from either an associated street or special building alarm, the second section is *not* to respond
(B) either a street alarm box or special building alarm, and shortly thereafter an alarm is received from an associated street or special building alarm, the second section is to respond
(C) street alarm box and shortly thereafter an alarm is received from an associated special building box, the second section is *not* to respond
(D) special building alarm and shortly thereafter an alarm is received from an associated street alarm box, the second section is *not* to respond.

14. "The fireman assigned to Housewatch Duty is responsible for giving and receiving alarms, not all of which come by way of telegraph signals." Suppose that, while you are on duty, a woman runs up to you. Breathless and almost hysterical, she shouts "Fire!" Of the following, the most important question for you, as House Watchman, to ask is
(A) "What is your name and address?"
(B) "Where is the fire?"
(C) "How serious is the fire?"
(D) "Did you actually see the fire yourself?"
(E) "How far is the fire from here?"

Key Answers: Alarm Signals

1. A	4. C	7. D	10. D	13. D
2. B-C	5. E	8. C	11. B	14. B
3. D	6. E	9. C	12. A	

RESCUE

1. The one of the following which is not a recommended procedure when administering artificial

respiration to another fireman is to
(A) continue the process without interruption
(B) keep a regular and definite rhythm
(C) keep the patient's nose and mouth free for breathing
(D) loosen tight clothing about the patient's neck and chest
(E) administer liquid stimulants at frequent intervals.

2. As your company arrives at the scene of a fire in a large rooming house, a man is spied on the second story fire escape about to let himself down to the ground by means of a sheet which he has torn into strips, knotted, and attached to some object in the building. Two middle aged women and a young boy are observed descending to the street by means of the fire escape. Flames and smoke are issuing from the third floor windows. Of the following, the best characterization of this man's behavior is that it is
(A) intelligent, because the fire has cut off egress by means of the stairway
(B) unintelligent, because a firm anchor for the knotted sheet could be achieved by securing it to the fire escape
(C) intelligent, because the fire may heat the fire escape to the point where it is red hot
(D) unintelligent, because to descend the fire escape would be a quicker and safer means of escape
(E) intelligent, because it has often been demonstrated that a properly knotted sheet is sufficiently strong to bear the weight of the average man.

3. When rescue operations are necessary at an airplane crash fire, certain methods are desirble. Of the following, the one which is *not* acceptable is that
(A) normally, two rescue men should be used for rescue operations at any single point of attempted entry.
(B) no part of the airplane structure should be moved unless absolutely essential to rescue operations
(C) not more than one man in a rescue team should make an initial entry into an airplane
(D) if cutting is required, usually three sides of a rectangular opening large enough to permit access will have to be fully cut
(E) cuts in the fuselage should be so laid out as to cut between two or more closely spaced rows of rivets.

4. The cause of the majority of fire deaths is
(A) shock
(B) asphyxiation
(C) burns
(D) infection
(E) falling material.

5. When a fire spreads to the stairs, elevators and corridors of an old non-fireproof hotel the greatest life hazard is usually found
(A) in collapse of interior stairs at lower floor levels
(B) on the top floor
(C) in places of public assembly above the ground floor
(D) in cellars and sub-cellars
(E) in storage rooms on floors above the ground floor.

6. Of the following, the one which is an acceptable practice in using an Atlas life net is that
(A) the maximum number of men that can efficiently man a net is eighteen
(B) the net should generally be held at the level of the neck with the right foot advanced in front of the left
(C) care be taken to see that the elbows of the net holders clear their bodies when the jumper lands
(D) the net should, when moving is necessary, be moved in a counterclockwise direction
(E) the minimum number of men required to hold the net is seven when the height to to be jumped is five floors.

7. Rescue work at a fire on the twelfth floor of a twelve story building can usually best be done by use of
(A) aerial and comb ladders
(B) pompier ladders
(C) stairways and fire escapes
(D) life line and belts.

8. Which of the following does not apply to handling life lines?
(A) store it so as to avoid distortion
(B) coil it clockwise in a bag

(C) allow one foot of the line to protrude from the life line bag
(D) throw it clear of building after use on roof.

9. All of the following are sound rescue operations at a fire except one. Indicate that one:
(A) care must be taken to prevent rescued persons from re-entering a burning building to secure personal property
(B) fire-fighting operations should be subordinate to rescue unless such operations are necessary to protect the lives of occupants and to insure their escape
(C) precaution should be taken to prevent untrained persons from exposing themselves to hazardous conditions
(D) age, sex, and physical conditions of occupants are of minor importance in setting up rescue operations.

10. "Firemen holding a life net should keep their eyes on the person jumping from a burning building." Of the following, the best justification for this recommendation is that
(A) a person attempting to jump into a life net may overestimate the distance of the net from the building
(B) some persons will not jump into a life net unless given confidence
(C) a person jumping into a life net may be seriously injured if the net is not allowed to "give" slightly at the moment of impact
(D) firemen holding a life net should be evenly spaced around the net at the moment of impact in order to distribute the shock
(E) an open net is suggestive and leads some people to jump unnecessarily.

Rescue

Key Answers

1. E	4. B	7. C	10. A
2. D	5. B	8. D	
3. E	6. C	9. D	

SIZE UP

1. When a fire is visible from the street at the front windows of a building, it is very frequently reasonable to assume that the fire has originated close to the windows and that a very small portion of the building or floor is involved. In such instances, the more efficient attack is from the
(A) outside of the building primarily because it is a more direct method of attack
(B) inside of the building primarily because streams can be pointed directly at the fire
(C) outside of the building primarily because heavier streams can be brought to bear on the fire
(D) inside of the building primarily because the hazards of an outside attack in varying weather conditions can be avoided
(E) outside of the building primarily because a speedier attack can be made in this way.

2. Perhaps the most important factor of a building that should be considered by the commanding officer in sizing up a fire is the
(A) occupancy
(B) construction
(C) height
(D) elevators and other shafts.

3. Following the size-up, it is better for the Fire Officer in charge to call
(A) a small amount rather than too much additional apparatus
(B) too much rather than too little apparatus
(C) just enough apparatus, which is usually possible to do
(D) sound one additional alarm and leave the remainder to the superior officer that will respond.

4. Suppose that you are a fireman and that your company has responded to a fire at a large apartment house. You are on a ladder, about to enter a smoke-filled room through the window in order to search for a woman reported to be still in the burning building. Of the following, the best action for you to take first before you enter the room is to
(A) make certain that there is at least one other means of exit from the room in the event of an emergency
(B) lash the ladder securely to a heavy piece of furniture in the room
(C) note the layout of the room in relation to all stairways in the building
(D) test the flooring around the window cau-

tiously with one foot to determine whether the flooring will hold your weight
(E) determine quickly whether the fire is spreading through the walls of the building.

5. Assume that you are a fireman fighting a fire on the third story of a six story building. Generally, it can be expected that the heat from this fire will spread
(A) upwards
(B) downwards
(C) sideways along the joists
(D) sideways along the walls
(E) evenly in all directions.

6. If during a fire it is observed that the water draining away from the fire is cold, it may be concluded that
(A) the water has played a negligible part in putting out the fire
(B) the hoses are placed to the windward
(C) the heat waves are forcing the water streams away from the fire
(D) more hose lines are needed to fight the fire
(E) the water pressure is too low.

7. Suppose that, in a fire in an unfinished warehouse, several iron and steel structural supports are heated to a point where they are red hot. Of the following, the best reason for not directing water against the heated supports is that
(A) the action of water on heated metal may produce a sudden burst of flame
(B) iron and steel are very poor conductors of heat
(C) hot sparks may be produced, causing the fire to spread to other parts of the warehouse
(D) smoke may be given off in excessively large quantities
(E) iron and steel supports may buckle when cooled quickly, causing the roof to collapse.

8. Suppose that a fire has occurred in a drug store basement where a large stock of chemicals is stored on open shelves and in wooden boxes. As a member of an engine company, you take a line of hose down into the basement. The smoke is very thick, but flames can be seen behind a row of boxes. Of the following, the chief precaution for you to take is to avoid
(A) upsetting and mixing the chemicals, lest there be an explosion
(B) wetting any wooden boxes near the flames, lest a chemical reaction result from the effect of water on wood
(C) wetting the walls, lest a short circuit be caused in the electric wiring
(D) opening any windows in the basement, lest the smoke be allowed to escape
(E) throwing water directly on the fire, lest excessive smoke be produced.

9. "Keep in mind that floor boards are, as a rule, laid lengthwise in a room." This advice can be most helpful to a fireman in
(A) keeping the amount of hose necessary to fight a fire at a minimum
(B) locating the path of a fire spreading within the walls of a room
(C) escaping from a room filled with dense smoke
(D) avoiding unnecessary damage to property while extinguishing a fire
(E) determining the exact location of a fire hidden by thick smoke.

10. Upon arrival of first due engine company at the front door of a 3 story brick rooming house, the officer in charge is confronted by a civilian who says, "Don't bring the hose in here, the fire is out. A mattress was burning in a room on the 3rd floor and they threw it out the window to the back alley." Which one of the following actions should the officer take?
(A) proceed with small line to 3rd floor room where fire started
(B) after directing the small line man to take the small line through the building to alley and extinguish the fire in mattress, proceed to 3rd floor room where fire started
(C) proceed with small line, through the building to the alley and extinguish fire in mattress
(D) after directing small line man to wait outside with the line, proceed to 3rd floor room where fire started
(E) after directing wagon driver to drive wagon around back and extinguish the fire in the mattress, proceed to 3rd floor room where fire started.

11. An alarm has been sounded for a fire in an electric substation. On arrival it is noted that large quantities of insulating oil are involved in fire. Which of the following gases presents the chief danger in such a fire?
(A) hydrogen cyanide (B) hydro carbon
(C) nitrogen (D) phosgene.

12. The officer in charge of operations at a fire has the responsibility for "sizing up" or evaluating the fire situation. Of the following factors, the

one which would have *least* influence on the "size up" is the
(A) time of fire
(B) contents of the building on fire
(C) insurance coverage
(D) amount of smoke
(E) height of the building on fire.

13. When your company arrives first at a fire in a large warehouse, you discover heavy smoke coming from one of the trailer trucks loaded and hooked up for the next day's deliveries. Of the following, the best action for you to take first is to
(A) attack the fire with hand lines
(B) open the trailer to find out what it contains
(C) apply small streams to the gasoline tank
(D) move the trailer away from the building.

14. Upon arriving at a fire in your fire district in a hospital building which was constructed for this use four years ago, you find the fire burning rather briskly in a room 20 by 30 feet, used for fumigation of bedding and furniture. This room has a separate stack and is fire resisting with no entrance from the inside. The fire door that leads to this room from the outside is tight, and the only escape for smoke and fumes is through the stack provided for the purpose which runs 20 feet above the roof of the hospital which is four stories high. This room has no other use. You should
(A) have several charged lines made ready, open the door and quickly flood the room
(B) make a small opening in the outer wall and direct several hose streams through this to extinguish the fire
(C) stand by and let the fire burn itself out. Have hose lines made ready for emergency
(D) have patients removed from this portion of the building, and attack the fire in the usual manner.

15. The phase of fire operations which most nearly corresponds to the "reconnaissance" element of military operations is
(A) routine inspection (B) size-up
(C) protection of exposures (D) salvage
(E) overhaul.

16. When the first due engine company arrives at a fire in an old-law tenement and it is found that the cellar is involved, the *first* step of the following should be to
(A) open the cellar door to ventilate it, and attack the fire from the rear stairs
(B) open roof skylight before opening the cellar door
(C) operate one line from the interior stairs into the cellar and a second line from the exterior stairs into the cellar
(D) use cellar pipes and distributors immediately to bring the fire under control
(E) take the first hose line into the main hallway to protect the interior stairway.

17. A mercantile building is filled with smoke but no fire is showing. The first company to arrive should check for the source of the fire
(A) in the vertical arteries and channels
(B) in any area where a fire might smoulder for a long time before being discovered
(C) in the basement
(D) on floors used for storage purposes
(E) on the upper floors.

Suppose that you are the first officer to arrive on the first alarm at a fire in a bedroom in the rear of the third floor of a four-story apartment house situated in a low pressure district. It is 8:30 P.M. The fire remains confined to this bedroom, but the contents of the room are totally consumed by the fire. There is only one elevator and one stairway in the building, which is equipped with a dry standpipe system. The elevator is out of order. There is some smoke in the third floor hallway.

Answer questions 18 to 20 on the basis of this situation.

18. To order that a second alarm be turned in immediately would be
(A) desirable; there is a considerable life hazard
(B) undesirable; a chief officer has not yet arrived
(C) desirable; the fire is relatively inaccessible
(D) undesirable; the fire is confined
(E) desirable; the non-operating elevator poses a serious problem.

19. Of the following, the best action for the first due engine company to take first is to
(A) carry a length of hose to the fourth floor standpipe outlet
(B) stretch a hand line to a point in the street from which the fire can be reached by a good stream
(C) lead a line from the pumping engine to the standpipe inlet
(D) connect a line to the hydrant and stretch it directly into the building
(E) wet down the stairway.

20. Suppose that the firemen were unable to ascend the stairs because of heat and smoke. Of the following, the best single method of ventilation to relieve this condition is to open the
(A) roof directly over the burning room
(B) windows along the elevator shaft
(C) windows on the third floor
(D) roof directly over the stairway
(E) windows at the point where the smoke and heat are most intense.

21. To order that the door of the room on fire be opened before a hose line is in position would be
(A) undesirable, because the fire may spread
(B) desirable, because the pressure of smoke and heat in the room would be relieved
(C) desirable, because no other means of ventilation is readily available
(D) undesirable because opening the roof would be a superior means of ventilation
(E) undesirable because removal of occupants from the building may still be in process.

22. Of the following, the most important precautionary action to be taken first is to
(A) cut a hole in the roof for ventilation
(B) shut off the gas supply to the building
(C) set up a deck gun and establish a water curtain
(D) open all the windows on the street level
(E) open a life net in front of the building.

23. Suppose that you notice five truckmen from the first due Hook and Ladder company entering the building. Three men are carrying axes, one a claw tool, and one a hook. The thought should immediately occur to you that this distribution of tools demonstrates poor judgment because
(A) more axes will probably be needed to open floor boards
(B) not all tools are needed at a particular fire
(C) more claw tools will probably be needed for overhauling furniture
(D) approximately an equal number of tools of different types should be carried
(E) more hooks will probably be needed for work on lath and plaster ceilings.

24. Of the following, the best procedure for extinguishing the fire is to
(A) flood the room with a heavy stream of water
(B) use fire extinguishers rather than water
(C) use a small tip on a charged hand line
(D) throw water simultaneously on the fire from the hallway and from an adjacent building
(E) cut off the supply of oxygen from the fire.

25. Suppose that you are the Lieutenant in command of a Hook and Ladder Company. When your company reaches a fire, first due, you see smoke issuing from all the windows of several floors. When you enter the building, there appears to be smoke on all floors, but no flame can be seen. Of the following, the first thought that should occur to you is
(A) "back draught"
(B) "cellar fire"
(C) "electrical short circuit"
(D) "defective sprinkler system"
(E) "carbon monoxide gas."

26. Suppose that you are the lieutenant commanding the first due engine company and respond to an alarm sounded for report of fire in a 5-story apartment building, approximately 50 feet by 100 feet in area. You lay out a 2½" line from the hydrant to the wagon, but you can see no immediate evidence of the location of the fire. Of the following, the best action for you to take under these circumstances is to
(A) order the 350′ preconnected 1½" line brought into the building
(B) endeavor to quickly locate the fire before ordering a specific line
(C) order the preconnected 3/4" line brought into the building
(D) order the 150′ preconnected 1½" line and the 1½" standpipe roll brought into the building.

27. In sizing up an exposure situation on the fire ground, the facts must be considered and probabilities estimated. Of the following, which one would most nearly be considered as a probability and not a fact:
(A) proximity of exposure
(B) nature and physical arrangement of involved and unconsumed fuel
(C) nature, volume and physical arrangement of exposed fuel
(D) ignition of expose by brands and sparks.

Size Up

Correct Answers

1. B	8. A	15. A, B	22. A, B
2. A	9. C	16. E	23. E
3. B	10. A	17. C	24. C
4. D	11. B	18. D	25. B
5. A	12. C	19. C	26. B
6. A	13. D	20. D	27. D
7. E	14. C	21. A	

VENTILATION

1 Suppose that your company is operating at a fire in a three-story factory. Your commanding officer orders you to open the skylight on the roof to ventilate the building. Openings for ventilation are usually made in the roof of a burning building chiefly because
   (A) hot air tends to flow towards the point of greatest pressure
   (B) air is much lighter than smoke
   (C) a rush of fresh air might cause the fire to spread
   (D) smoke tends to remain close to the seat of the fire
   (E) hot gases and smoke tend to rise.

2. Water, whether in solid or fog form, should not be used on a glowing or blazing fire in a poorly ventilated coal bunker mainly because
   (A) an explosive gas may be produced by such contact
   (B) unless well ventilated, excessive pressures may result from the enormous production of steam
   (C) the fire area will become so obscured by steam that effective follow-up will be impossible
   (D) the metal walls of the bunkers may crack and spread the burning coals
   (E) a violent up draft will result from the formation of steam on the coal surface.

3. Before attacking a fire in a closed building, it is necessary to determine whether ventilating is necessary and, if so, the method of ventilating. Which of the following would be of *least value* in making such decision
   (A) observing the points at which smoke may be oozing out from the building
   (B) observing density of smoke through windows, if possible
   (C) feeling walls and roof for hot spots
   (D) observing the color of the smoke

4. The following statements refer to ventilation practices. The one which is an incorrect practice is
   (A) men making a roof opening should stand on the windward side of the opening
   (B) in cross-ventilation, windows on the leeward side of the building should be opened first
   (C) in cross-ventilation, floors are cleaned of smoke and gases one at a time, working upward from the lowest floor involved
   (D) if properly located, natural opening such as skylights and scuttle holes should be used for vertical ventilation.

5. "Firemen operating on hot fires in the open have experienced the sensation of a cold draft on their back, while their faces were being scorched." The most likely reason for this phenomenon is the
   (A) excessively windy weather at the time of the fire
   (B) air movement toward the center of the fire
   (C) excessive heat at the fire periphery
   (D) improper placement of forces fighting the fire
   (E) high pressure areas created at the center of the fire.

6. The safety feature in the new type open air parking garage is
   (A) ventilation
   (B) moderate exposure
   (C) no heating equipment would be required
   (D) accessibility.

7. There can be no fixed rule for "opening" or ventilating at a fire. However, it is correct to assume that
   (A) openings are best made below the fire
   (B) openings are best made on the floor of fire
   (C) in tall buildings, roof openings are preferable procedure
   (D) in ordinary dwellings, especially of frame construction, smoke is seldom a material factor.

8. When ventilating a peaked roof, the firefighter should
   (A) make an opening on windward side of roof peak
   (B) rip off roof covering on lee side and cut sheathing
   (C) make opening at right angles to ridge pole
   (D) make openings parallel to ridge pole on windward side of roof.

9. "It takes at least two openings to do an effective job of horizontal or lateral ventilation." This statement is, in general,
   (A) correct; there must be an air inlet as well as an air outlet
   (B) not correct; the pressure of the hot gases tends to force them from the building once an outlet is made
   (C) correct; fire tends to follow through an open window if only one opening is made
   (D) not correct; the number of openings should be kept at a minimum in order to keep

down the number of exposures
(E) of no importance; the number of openings depends on the particular case and cannot be determined in advance.

10. The one of the following which is *least* desirable as a practice in cutting a hole in a roof is to
(A) cut at an angle to the grain of the wood instead of straight across it
(B) use short strokes rather than long strokes when using an axe
(C) cut diagonal sheathing along the studding in the direction toward which it runs
(D) stand on the leeward side of the boards to be cut
(E) cut as close as possible to a beam or joist.

11. When an opening has been made in a roof solely for the purpose of ventilation, it is unwise to play water into the hold primarily because the
(A) hazard to the men operating on the roof will be increased
(B) water will tend to drive the gases inward
(C) men operating below will suffer badly from the water
(D) hot gases will be cooled so that they will not rise
(E) determination of the need for more ventilation will be made difficult.

12. "The smoke in the basement was very thick. The basement was ventilated by tearing up a board in the floor above the basement and opening windows at the front and rear of the building." Of the following, the chief precaution to be observed when ventilating in this manner is to make certain that
(A) a Bresnan distributor or cellar pipe is already in action
(B) the fire covers a relatively small part of the cellar
(C) a hose line is in position near the floor opening
(D) the stairway into the basement is free of encumbrances
(E) the floor opening is directly over the fire.

13. Suppose that your company has responded to a fire alarm at a two-story laundry. The commanding officer, after noting quickly the extent to which the fire has taken hold, orders you to the roof to open the skylights as quickly as possible. Of the following, the best reason for this command is that it is probably necessary to
(A) admit a sufficient supply of oxygen to the fire
(B) prevent collapse of the roof by reducing atmospheric pressure
(C) fight the fire from below rather than above
(D) gain access to the building in this manner
(E) prevent the accumulation of gas and heat under the roof.

14. Ventilation is a subject that has been much discussed by architects and engineers. Differences of opinion as to the best method of ventilating a school building (for normal usage) effectively are so pronounced as to make this almost a controversial matter. However, in point of fire hazard, which of the following ways is the agreed safer design?
(A) from the roof down;
(B) opening the windows ⅔ from top and ⅓ from bottom;
(C) mechanical means of ventilation;
(D) natural ventilation.

15. It frequently occurs that an engine company will go up stairways and have water at hand before the truck company has had an opportunity to ventilate. It may be hot, smoky, and impossible to make the fire floor under the existing conditions. The engine company, under such conditions, should
(A) adjust the nozzle for fog to cool the surroundings and advance cautiously to the fire area
(B) have another engine company back you up, both with nozzles adjusted for fog, and advance cautiously to fire area
(C) avoid the use of water until ventilation is made
(D) open a solid stream into fire area as truck company ventilates the building.

16. "Good ventilation practices require careful analysis of the situation as well as a basic knowledge of hostile fire behavior." Of the following, the question which best illustrates the most comprehensive understanding of the purposes and value of ventilation is:
(A) will ventilation at this point tend to spread the fire
(B) are smoke and hot gases being generated by the fire
(C) should this window be opened or smashed
(D) will ventilation at this point cause damage
(E) should this door be forced or smashed open.

17. Which one of the following statements about the use of aerial ladders for knocking out windows for ventilation on the fire ground is correct?

(A) the fluid lock valve must be open when punching out windows with the fly section
(B) jacks must be down and in place
(C) the truck must be jack-knifed in the direction of the opening being ventilated
(D) the aerial ladder must never be extended more than 80 ft. when punching out windows.

18. Back draft is most likely to occur at a fire in a building charged with fumes and smoke and in which the fire has been burning for some time and is fairly well closed in. Of the following, the first sign of a probable back draft is that, as soon as doors are opened
(A) flames will be seen to build up gradually
(B) no flames can be seen
(C) air is felt to rush out of the building
(D) the smoke lies low along the ground
(E) the smoke seems to be drawn into the building sharply.

19. Assume that there is a considerable fire contained in a cellar below grade, having no outside vent or entrance, in a retail shoe store located in a five story brick building. With regard to immediate fire-fighting procedures, which of the following is of least consequence?
(A) ventilation of the first floor
(B) ventilation at the roof
(C) protection of interior exposures
(D) use of 2½" hose lines
(E) cutting holes in floor on the first floor.

20. Suppose that, after cutting a hole in the roof of a fire building, you find that only a small draft is set up. In this situation, the first action that should be taken is to
(A) probe the hole with a pole for obstructions
(B) enlarge the opening
(C) see whether the smoke is escaping from another opening
(D) start a second opening at a spot some distance away.

21. When companies respond to an alarm for a building where fumigation with hydrocyanic acid gas is being conducted, it is imperative that thorough and complete ventilation be effected immediately, primarily because in the concentration usually used
(A) rapid corrosion of protective devices occurs
(B) application of water or foam is of limited effectiveness
(C) the gas is explosive even at room temperature
(D) the gas is absorbed through unprotected skin

22. The one of the following gases which is lighter than air and therefore requires ceiling ventilation is
(A) benzol (B) ammonia
(C) hydrogen sulphide (D) sulphur dioxide
(E) nitric acid fumes.

Key Answers: Ventilation

1. E	6. A	11. B	16. A	21. D
2. A	7. C	12. C	17. B	22. B
3. D	8. B	13. E	18. E	
4. C	9. A	14. D	19. B	
5. B	10. D	15. C	20. A	

EXPLOSIVE HAZARDS

1. An intense fire develops in a room in which carbon dioxide cylinders are stored. The principal hazard in this situation is that
(A) the CO_2 may catch fire
(B) toxic fumes may be released
(C) the cylinders may explode
(D) released CO_2 may intensify the fire.

2. The explosive force of a gas when stored under various pressures is given in the following table:

Storage Pressure	Explosive Force
10	1
20	8
30	27
40	64
50	125

The one of the following statements which best expresses the relationship between the storage pressure and explosive force is that
(A) there is no systematic relationship between an increase in storage pressure and an increase in explosive force
(B) the explosive force varies as the square of the pressure
(C) the explosive force varies as the cube of the pressure
(D) the explosive force varies as the fourth power of the pressure.

3. Suppose that a building in which a fire has occurred has been used to store cylinders of compressed carbon dioxide gas. It is important

that you cool the cylinders with large quantities of water because the
(A) gas may become ignited and cause serious damage
(B) metal cylinders may melt and release the gas
(C) heat may cause the gas to condense and form hydrochloric acid
(D) heat may increase the pressure of the gas to the point that the cylinders will burst
(E) rapid decompression of the gas due to the heat, in the presence of oxygen from the air, may cause an explosion.

4. Electric blasting caps are particularly susceptible to accidental detonation by which of the following:
(A) X-ray machines
(B) direct current generators
(C) radio transmitters
(D) fluorescent lighting
(E) ultraviolet light.

5. In the event of an atomic attack a fire department trained to conduct its own radiological monitoring should be able to accomplish all of the following objectives except one:
(A) minimize diversion of regular monitoring crews
(B) enable the officer in charge to make better decisions involving calculated risks
(C) dangerous areas could be outlined as firemen proceed with advance fire-fighting activities
(D) enable fire fighters to promptly function with confidence whether or not the radiological area monitor is present at any particular moment.

6. A fire service monitoring program should consider all of the following items of equipment except one. Indicate that one:
(A) sensitive radioactive instruments are required in the initial operations
(B) high level measuring instruments should be provided on the basis of one for each fire house
(C) some type of direct reading dosemeter should be provided for every officer to measure accumulated dosage
(D) a film badge or other device should be provided for each individual fire fighter to give a permanent record of his exposure and so guide any medical treatment necessary.

7. Of the following statements concerning the explosion hazard of dust, the one that is *least* correct is that
(A) dust explosions frequently occur in pairs
(B) there are explosive limits beyond which an explosion cannot occur
(C) the likelihood of an explosion is affected by particle size and amount of impurities present
(D) the intensity of an explosion is affected by the duration of a source of ignition.

8. Of the following conditions, the one which is most favorable for an explosion of flammable vapors is
(A) high humidity
(B) high diffusion rate
(C) low igniting temperature
(D) low volatility
(E) much air movement.

9. "Never extinguish a gas fire, unless by extinguishing it you can gain access to the valve to shut off the flow." This statement is
(A) correct; in fact it is not possible to extinguish such a fire without shutting off the valve
(B) not correct; it is important to extinguish the fire as soon as possible
(C) correct; if gas continues to escape after the flames are extinguished there will be the possibility of an explosion
(D) not correct; the heat produced by the flame is sufficient to raise additional fuel above its flash point and cause an extension
(E) correct; extinguishing the fire may make it impossible to shut off the valve as the pressure in the line will tend to increase.

10. Pouring flammable liquids from one container to another presents fire hazards. The most common fire hazard is the possibility of
(A) explosion caused by a static spark
(B) spontaneous combustion
(C) fire caused by the heat friction
(D) the liquid vaporizing and leaving an explosive pocket
(E) a flash fire resulting from the presence of an open flame.

11. The greatest practical difference between dust explosion and a gas explosion is that generally in the case of gas explosions
(A) the damage done is greater than a dust explosion

(B) it is harder to predict than a dust explosion
(C) the greatest damage is found at some point far from where the explosion started
(D) the force of the explosion is greater in the vicinity of origin.

12. Responsibility for the majority of anesthetic explosions in hospital operating rooms is usually credited to
(A) electric hot plates and similar heating devices
(B) defective anesthetic equipment controls
(C) smoking in the operating room area
(D) improper control of air or oxygen flow in the room
(E) electrostatic discharges.

13. The one of the following substances which has the greatest explosion hazard is
(A) leather dust (B) alfalfa
(C) linseed meal (D) wood dust
(E) paper dust.

14. The *least* effective solution to eliminate possible smoke explosions in a building is
(A) tight floors
(B) protection of vertical openings
(C) ventilation
(D) wire-glass windows.

15. The modern explanation for "dust explosions" is that such explosions occur when
(A) certain combinations of gas and dust form the proper mixture
(B) dust of certain substances, properly dispersed, is ignited by a source of heat
(C) gas explodes in the presence of dust
(D) dust when heated liberates gas, which in turn, explodes.

16. The best precaution in preventing dust explosions is
(A) dust-tight equipment
(B) a dust collection system
(C) education of personnel as to explosion hazards
(D) cleanliness
(E) venting.

17. Where explosive dust and flyings exist, all electrical installations and equipment shall comply with the requirements of the
(A) Board of Fire Commissioners
(B) National Fire Protection Association
(C) Bureau of Highways
(D) Building Code.

18. Explosions of vapor-air mixtures inside tanks containing flammable liquids are most frequent when the tank is
(A) full
(B) has just been emptied
(C) is being filled
(D) is half full.

19. Flammable gases and vapors will burn or explode
(A) only when the proportion of gas or vapor to air is within certain maximum or minimum limits
(B) when mixed with air in any proportion
(C) when mixed with air in a proportion down to and not below a certain amount
(D) when mixed with air in any proportion up to but not above a certain amount.

20. Suppose that a fire has occurred in a drugstore basement where a large stock of chemicals are stored on open shelves and in wooden boxes. As a member of an engine company, a fire fighter takes a line of hose down into the basement. The smoke is very thick, but flames can be seen behind a row of boxes. Of the following, the chief precaution for the fire fighter to take is to avoid
(A) upsetting and mixing the chemicals, lest there be an explosion
(B) wetting any wooden boxes near the flames, lest a chemical reaction result from the effect of water on wood
(C) wetting the walls, lest a short circuit be caused in the electric wiring
(D) opening any windows in the basement, lest the smoke be allowed to escape.

21. Tests on dust explosion hazards of powders show that as the size of the particle increases the ignition temperature
(A) increases and the rate of pressure rise during the explosion is decreased
(B) and the rate of pressure rise during the explosion both increase
(C) and the rate of pressure rise during the explosion remain relatively unchanged
(D) and the rate of pressure rise during the explosion both decrease
(E) decreases and the rate of pressure rise during the explosion is increased.

22. The hazard of static sparks can *least* be avoided by

    (A) neutralizing (B) humidification
    (C) grounding (D) isolation.

Key Answers: Explosive Hazards

1. C	6. A	11. D	16. D	21. A
2. C	7. D	12. E	17. D	22. D
3. D	8. C	13. D	18. B	
4. C	9. C	14. D	19. A	
5. C	10. A	15. B	20. A	

CHEMISTRY AND PHYSICS FOR ENTERING FIREMEN

"Knowledge of the principal properties of everyday chemicals is a part of the stock in trade of every scientific fire fighter." Column I below lists ten chemicals. Column II lists four descriptive statements. On your Answer Sheet, next to the number for each of the chemicals in Column I, write the letter of the one statement in Column II which best describes that chemical.

COLUMN I	COLUMN II
1. carbon dioxide	(A) a gas at ordinary temperatures which has a characteristic color or odor
2. hydrogen	(B) a gas at ordinary temperatures which is lighter than air
3. nitric acid	(C) a gas at ordinary temperatures which is non-inflammable
4. carbon tetrachloride	(D) a chemical which is not a gas at ordinary temperature
5. methane	
6. sulphur	
7. carbon monoxide	
8. ethyl chloride	
9. hydrogen sulphide	
10. sodium chloride	

11. "Before entering a room filled with carbon monoxide, tie a wet cloth over your mouth and nose." The procedure described in the above quotation
    (A) is useless
    (B) is effective only for preventing asphyxiation
    (C) is valuable for no longer than 5 minutes, by which time the water will have evaporated from the handkerchief
    (D) will furnish protection against carbon monoxide but not against carbon dioxide
    (E) will be effective as long as the handkerchief remains moist.

12. Of the following, the best conductor of heat is
    (A) concrete (B) porcelain
    (C) pine wood (D) steel
    (E) bricks.

13. "Water possesses certain advantages as an extinguishing agent." Of the following, the chief advantage of water as an extinguishing agent is that water
    (A) enters into chemical union with certain substances, producing heat
    (B) has a high freezing point
    (C) is lighter than oil and will not mix with it
    (D) has a great capacity for absorbing heat
    (E) is decomposed, producing combustible gases, when in contact with hot metals.

14. A tension pulley properly used on a belt connecting two pulleys of different size
    (A) is generally placed nearer the large pulley
    (B) is generally placed nearer the small pulley
    (C) in no way adds to the belt life
    (D) will commonly be a crowned tension pulley.

15. The one of the following statements about smoke which is most accurate is that smoke is
    (A) irritating but not dangerous in itself
    (B) irritating and dangerous only because it may reduce the oxygen content of the air breathed
    (C) dangerous because it may reduce the oxygen content of the air breathed and often contains toxic gases
    (D) dangerous because it supports combustion.

16. If boiling water is poured into a drinking glass, the glass is likely to crack. If, however, a metal spoon first is placed in the glass, it is much less likely to crack. The reason that the glass with the spoon is less likely to crack is that the spoon
    (A) distributes the water over a larger surface of the glass
    (B) quickly absorbs heat from the water
    (C) reinforces the glass
    (D) reduces the amount of water which can be poured into the glass.

17. The one of the following substances with the *least* tendency to spontaneous heating is
    (A) fish meal (B) lamp black
    (C) scrap rubber (D) soap powder.

18. When radioactive materials are involved in hot fires, the radiation coming from the substance
    (A) is not affected
    (B) increases in rate and intensity
    (C) decreases in rate and intensity
    (D) increases in rate and decreases in intensity.

19. "A soft, wax-like metal, which rapidly decomposes water, on which it floats. It is inflammable and may explode spontaneously on contact with water. It is used in chemical laboratories and is kept under oil or paraffin in glass bottles, tin cans and iron drums." This description best applies to
    (A) picric acid (B) formic acid
    (C) nitro aniline (D) chloride of lime
    (E) potassium.

20. "It ignites organic matter and explodes when brought in contact with alcohol or acetic acid. It is a strong oxidizing agent and should be isolated in storage, for when involved in fire it may cause an explosion." This description best applies to
    (A) chromic acid (B) charcoal
    (C) celluloid (D) barium dioxide
    (E) borneol.

In column I are listed a number of substances. Column II gives four classifications of these substances at ordinary temperatures. For each substance in Column I, write A in the space on the left if the substance is a gas; if it is a liquid write B; if it is a metal write C; and if it is a non-metallic solid write D in the proper space.

COLUMN I	COLUMN II
21——carbon dioxide	(A) gas
22——sulphur acid	(B) liquid
23——nitrogen	(C) metal
24——suplhur	(D) non-metallic solid.
25——magnesium	
26——calcium	
27——carbon	
28——hydrogen chloride	
29——hydrogen oxide	
30——chlorine	

31. The one of the following which is not an irritating gas when heated is
    (A) sulphur dioxide (B) carbon monoxide
    (C) ammonia (D) bromine gas
    (E) vaporized nitric acid.

32. Elements of illuminating gas are
    (A) carbon and hydrogen
    (B) silicon and chlorine
    (C) sodium and potassium
    (D) oxygen and calcium
    (E) magnesium and phosphorus.

33. The one of the following which is a gas and is not inert is
    (A) neon (B) nitrogen
    (C) fluorine (D) iodine
    (E) bromine.

Chemistry and Physics

Correct Answers

1. C	8. A	15. C	22. B	29. B
2. B	9. A	16. B	23. A	30. A
3. D	10. D	17. B	24. D	31. B
4. D	11. A	18. A	25. C	32. A
5. B	12. D	19. E	26. C	33. C
6. D	13. D	20. A	27. D	
7. B	14. B	21. A	28. A	

HYDRAULICS FOR ENTERING FIREMEN

1. Firemen should possess an extensive knowledge of hydraulics mainly because
    (A) water is generated by hydraulic combustion
    (B) hydraulics is a compound of hydrogen and fluorine
    (C) hydraulics treats of water in motion
    (D) the study of gas involves a knowledge of hydraulic chemistry
    (E) a body has uniform motion over any specified distance.

2. It takes more energy to force water through a long pipe than through a short pipe of the

same diameter. The principal reason for this is
(A) gravity (B) friction
(C) inertia (D) cohesion.

3. In an experiment, a sprinkler system discharging W gallons of water per hour extinguished a fire covering a floor of A square yards in T minutes. The amount of water actually used to put out the fire was
(A) W times T divided by 60
(B) 60 times W divided by T
(C) 60 times W times T
(D) T divided by the fraction whose numerator is W and denominator 60
(E) none of the foregoing.

4. Suppose that the loss of water pressure in a hose due to friction is uniformly L pounds per square inch for every foot of hose. Of the following, the best estimate of the total loss in terms of pressure per square inch in a hose H feet long is
(A) H plus L pounds
(B) H times L pounds
(C) H divided by L pounds
(D) L divided by H pounds
(E) none of the foregoing.

5. When filled with water, a 50 foot section of a 2½" hose, assuming no expansion of the hose, contains water weighing most nearly
(A) 75 lbs. (B) 85 lbs.
(C) 95 lbs. (D) 105 lbs.

6. If the discharge from a 1¼" nozzle is 300 gallons per minute, the nozzle pressure is approximately
(A) 35 lbs. (B) 42 lbs.
(C) 49 lbs. (D) 56 lbs.

7. Suppose that it is necessary to compute the number of gallons of water contained in a length of hose 50 feet long and 2½ inches in diameter. Of the following, the chief additional item of information required is the
(A) radius of the hose
(B) circumference of the hose
(C) pressure of a gallon of water
(D) velocity of the water passing through the hose
(E) number of cubic inches in a gallon.

Figure A is a schematic drawing representing a Fire Department hose connection. The numbers 1 and 2 indicate valves. Inlet 3 is connected by means of a hose to one pumper, some distance away, and inlet 4 is connected to another pumper, equally distant from the connection. Outlet 5 leads to the nozzle from which water is thrown on the fire.

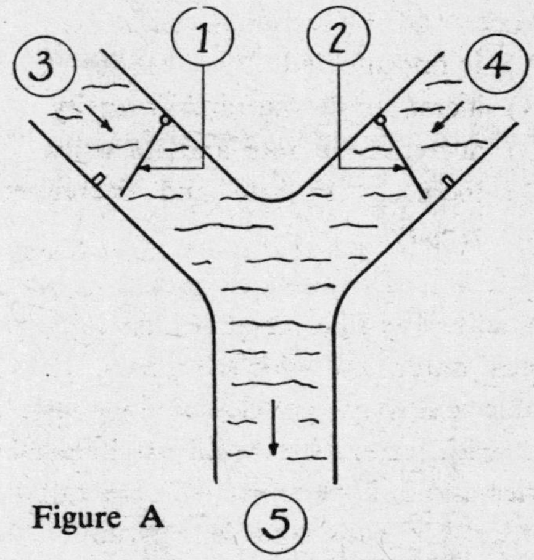

Figure A

Answer questions 8 to 10 on the basis of this drawing.

8. The chief purpose of valves 1 and 2 is to
(A) reduce the friction between the water and the connection
(B) increase the speed of the water in the connection
(C) prevent the pumpers from operating at excessive speeds
(D) prevent water from one pumper going back into the other pumper
(E) prevent excessive water pressure at the nozzle.

9. Suppose that both pumpers are operating and water is flowing through the nozzle. Valve 1 is open, but valve 2 is closed. Then, of the following, the most accurate statement is that the pressure at
(A) outlet 5 is greater than the pressure at inlet 3
(B) inlet 4 is greater than the pressure at outlet 5
(C) inlet 3 is greater than the pressure at inlet 4
(D) inlet 4 is equal to the pressure at inlet 3
(E) outlet 5 is equal to the pressure at inlet 4.

10. Suppose that the pressure at inlet 4 is much greater than the pressure at inlet 3. On the basis of this information, the most accurate statement is that
(A) valve 1 will be open and valve 2 will be closed
(B) valve 2 will be open and valve 1 will be closed
(C) valves 1 and 2 will both be open

(D) valves 1 and 2 will both be closed
(E) it is impossible to determine which valve will be open because no information is given concerning the distance of the nozzle from the connection.

Figure B is a schematic drawing of a pressure relief device employed on fire equipment. The number 1 indicates a piston having circular plates of equal size rigidly attached at its two ends. The piston may slide back and forth. Chambers 2 and 3 are filled with water under pressure. A pipe, 4, leads from chamber 2, as indicated.

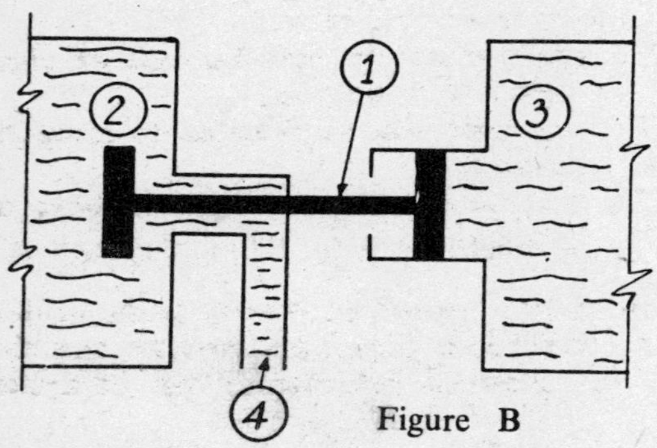

Figure B

Answer questions 11 to 13 on the basis of this drawing.

11. If the pressure of the water in chamber 2 is greater than the pressure of the water in chamber 3, the most likely result is that
(A) water will flow from pipe 4 into chamber 2
(B) water will flow from chamber 2 into chamber 3
(C) the water in pipe 4 will be compressed
(D) the piston will slide to the right as far as possible
(E) the piston will move back and forth regularly.

12. If the pressure of the water in chamber 3 rises suddenly to a point where it greatly exceeds the pressure of the water in chamber 2, the most likely result is that
(A) the piston will slide to the left as far as possible
(B) water will flow from chamber 3 into chamber 2
(C) the piston will slide back and forth regularly
(D) the water in chamber 2 will be compressed
(E) the piston will maintain a fixed position until the pressure in chamber 2 falls.

13. Assume that water is flowing out of chamber 2 through pipe 4. Then, of the following, the most accurate statement is that
(A) there can be no pressure in chamber 3
(B) there can be no pressure in chamber 2
(C) the piston has moved as far to the right as possible
(D) the pressure in pipe 4 must be higher than the pressure in chamber 3
(E) the pressure in chamber 3 must be higher than the pressure in chamber 2.

14. With an increase in the diameter of a pipe, the cross-sectional area
(A) increases at a lesser rate than the frictional resistance to flow
(B) decreases in proportion to the square of the radius
(C) decreases in proportion to the square root of the radius
(D) becomes equal to the circumference
(E) increases in greater proportion than the circumference.

15. "The frictional resistance of water pipes and consequent loss of pressure become relatively less as the size of the pipe is increased." It follows
(A) that the longer the pipe the more efficient is its performance
(B) that the loss of pressure in a given length of 2-inch pipe is greater than that occurring when the same amount of water flows through an equal length of 6-inch pipe
(C) pressure loss due to friction may be reduced by reducing the length of the pipe
(D) pressure in a length of 6-inch pipe is about half that of a 3-inch pipe
(E) the pressure in a length of pipe is equal to the reciprocal of its friction.

16. The usual rule, satisfactory for most applications, to find the discharge at a given pump pressure is
(A) divide the pressure at which the pump is rated by the rate discharge and multiply by the pressure at which you desire to find the discharge
(B) divide rated discharge by rated pressure
(C) multiply rated pressure by the reciprocal of the rated discharge
(D) multiply the pressure at which the pump is rated by the discharge at that pressure and divide by the pressure at which you desire to find the discharge

(E) multiply the rated pressure by the square of the rated discharge.

17. A tank is 10 feet high and 10 feet long. The pressure in pounds per square inch due to one foot elevation of water is .434. When the tank is full of water, the average pressure in pounds per square inch against the side of the tank is
(A) 31.248 (B) 2.17
(C) 43.40 (D) 4.34
(E) 8.68.

Answer items 18 to 22 on the basis of Figure C.

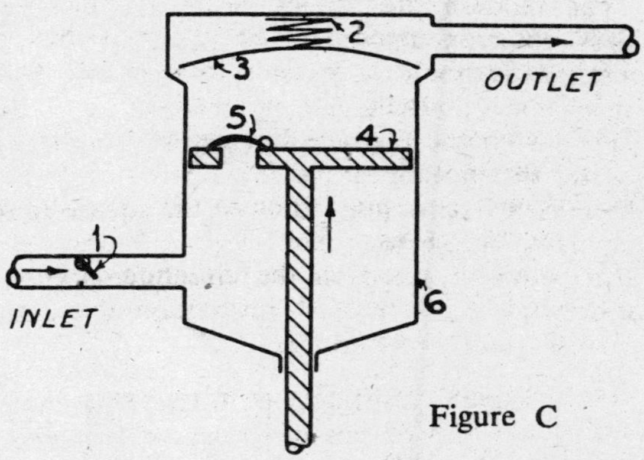

Figure C

The diagram shows a water pump in cross section. 1 is a check valve, 2 and 3 are the spring and diaphragm respectively of the discharge valve; 4 is the pump piston; 5 is the inlet valve and 6 is the pump cylinder. All valves permit the flow of water in one direction only.

18. When water is flowing through the outlet pipe
(A) check valve 1 is closed
(B) diaphragm 3 is closed
(C) valve 5 is closed
(D) the piston is on the downstroke.

19. If valve 5 does not work properly and stays closed, then
(A) the piston cannot move down
(B) the piston cannot move up
(C) diaphragm 3 cannot open
(D) check valve 1 cannot close
(E) the flow of water will be reversed.

20. If diaphragm 3 does not work properly and stays in the open position, then
(A) check valve 1 will not open
(B) valve 5 will not open
(C) spring 2 will be compressed
(D) spring 2 will be extended
(E) water will not flow through the inlet pipe.

21. When valve 5 is open during normal operation of the pump, then
(A) spring 2 is fully compressed
(B) the piston is on the upstroke
(C) water is flowing through check valve 1
(D) a vacuum is formed between the piston and the bottom of the cylinder
(E) diaphragm 3 is closed.

22. If check valve 1 jams and stays closed, then
(A) valve 5 will be open on both the upstroke and downstroke of the piston
(B) a vacuum will tend to form in the inlet pipe between the source of the water supply and check valve 1
(C) pressure on the cylinder side of check valve 1 will increase
(D) less force will be required to move the piston down
(E) more force will be required to move the piston down.

Below are two paragraphs relating to the work of firemen. Read each paragraph carefully and then answer the questions referring to that paragraph.

Paragraph 1

"Everyone knows that 'water seeks its own level'; that it will always flow to the lowest accessible point. If restrained, it exerts pressure against the restraining object, whether this be the walls of a drinking glass, a giant dam, or a pipe to conduct it from one place to another. The degree of pressure exerted upon the walls of a container at any point depends, not on the quantity of water stored, but on the vertical height to which it is backed up; in other words, the difference in elevation between the point where the pressure is measured and the surface of the continuous body of water that is restrained. The technical term for this difference of elevation is 'head', and in this country it is usually measured in feet. Firemen are more accustomed to thinking of water pressure in terms of pounds per square inch, usually abbreviated to 'pounds.' It is very easy to convert pressure in terms of 'head' to 'pounds'; simply multiply the head in feet by 0.433, and the result is pressure in terms of pounds per square inch."

Answer questions 23 to 29 on the basis of the information appearing in the above paragraph.

23. According to the above paragraph, the "head" at a point 4 feet above the bottom of a tank which is 12 feet deep and is filled with water to within 2 feet of the top of the tank is
(A) 4 feet (B) 6 feet
(C) 8 feet (D) 10 fect
(E) 12 feet.

24. The above paragraph explains what is meant by the principle "water seeks its own level." Of the following, the practice which is based most directly upon this principle is
   (A) using chemicals rather than water in fire extinguishers
   (B) placing sprinkler systems in basements rather than in the upper stories of a building
   (C) using 2½ inch hose for fighting fires rather than 3 inch hose
   (D) placing fire hydrants at frequent intervals along the street
   (E) placing a water tank on the roof of a building.

25. The number 0.433 mentioned in the above paragraph is best defined as the
   (A) temperature of water at a specific atmospheric pressure
   (B) area of the base of a specific restraining container
   (C) "head" at a point one foot below the surface of the water
   (D) weight of a specific quantity of water
   (E) quantity of water stored in a specific container.

26. According to the above paragraph, water pressure is zero
   (A) unless the "head" is measured in pounds per square inch
   (B) where the quantity of water stored is excessively large
   (C) when water is flowing through a fire hose
   (D) if the pipe conducting the water from one place to another is less than one square inch in diameter
   (E) at the surface of a large body of water restrained by a dam.

27. Suppose that water is confined in a tank. According to the above paragraph, pressure is exerted by the water
   (A) against the walls of the tank
   (B) in a direction level with the tank
   (C) only if the water seeks its own level
   (D) equivalent to atmospheric pressure
   (E) inversely with depth.

28. According to the above paragraph, to say that the water pressure at a given point is 200 pounds means most nearly that
   (A) the weight of all the water above the restraining point is 200 pounds
   (B) the force exerted by the water at that point is 20 pounds per square inch
   (C) the "head" is 200 pounds
   (D) a cubic foot of water at that point weighs 200 pounds
   (E) the vertical height of the water exceeds the equivalent of 200 pounds.

29. On the basis of the above paragraph, "head" may also be accurately expressed in
   (A) cubic feet
   (B) square inches
   (C) foot pounds
   (D) degrees
   (E) yards.

Paragraph 2

"The modern fire fighter is highly skilled and utilizes complex equipment in extinguishing fires. Consequently, the fireman must have a high degree of mechanical aptitude and he must also be well acquainted with the physical laws governing the flow of water through pipes and hose." Questions 91 to 100 present schematic drawings of fire fighting equipment. Read the explanation alongside each drawing and then answer the question relating to that drawing.

FIGURE I

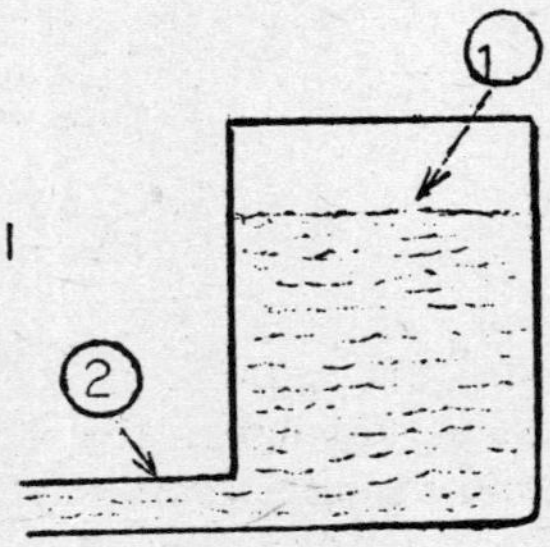

30. Figure I represents an enclosed water chamber, partially filled with water. The number 1 indicates air in the chamber and 2 indicates a pipe by which water enters the chamber. If the water pressure in the pipe, 2, increases then the
   (A) water pressure in the chamber will be decreased
   (B) water level in the chamber will fall
   (C) air in the chamber will be compressed
   (D) air in the chamber will expand
   (E) water will flow out of the chamber.

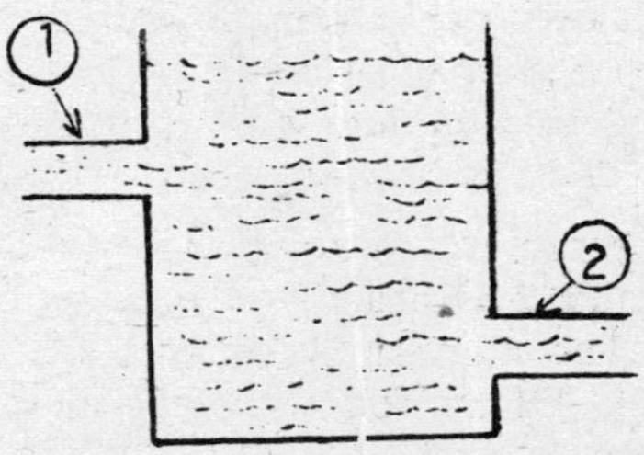

FIGURE II

31. Figure II represents a water tank containing water. The number 1 indicates an intake pipe and 2 indicates a discharge pipe. Of the following, the statement which is least accurate is that the
    (A) tank will eventually overflow if water flows through the intake pipe at a faster rate than it flows out through the discharge pipe
    (B) tank will empty completely if the intake pipe is allowed to remain open
    (C) water in the tank will remain at a constant level if the rate of intake is equal to the rate of discharge
    (D) water in the tank will rise if the intake pipe is operating when the discharge pipe is closed
    (E) time required to fill the tank, if the discharge pipe is closed depends upon the rate of flow of water thru the intake pipe.

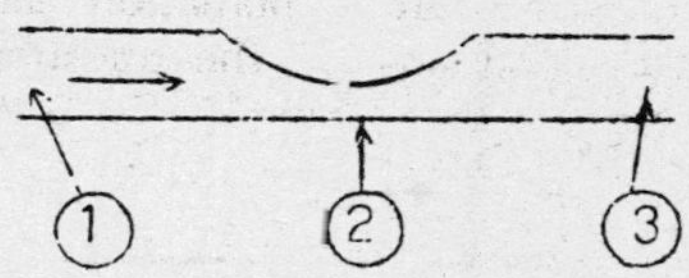

FIGURE III

32. Figure III represents a pipe through which water is flowing in the direction of the arrow. There is a construction in the pipe at the point indicated by the number 2. Water is being pumped into the pipe at a constant rate of 350 gallons per minute. Of the following the most accurate statement is that
    (A) the velocity of the water at point 2 is the same as the velocity of the water at point 3
    (B) a greater volume of water is flowing past point 1 in a minute than is flowing past point 2
    (C) the velocity of the water at point 1 is greater than the velocity at point 2
    (D) the volume of water flowing past point 2 in a minute is the same as the volume of water flowing past point 1 in a minute
    (E) a greater volume of water is flowing past point 3 in a minute than is flowing past point 2.

33. Figure IV represents a revolving wheel. The number 1 and 2 indicate two fixed points on the wheel. The number 3 indicates the center of the wheel. Of the following, the most accuate statement is that

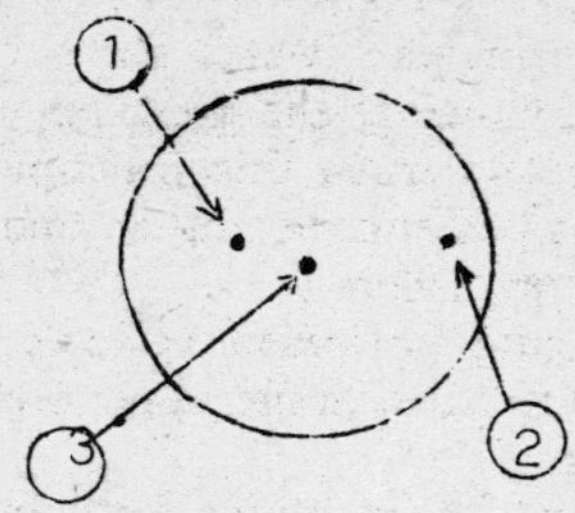

FIGURE IV

    (A) point 1 makes more revolutions per minute than point 2
    (B) point 2 makes more revolutions per minute than point 1
    (C) point 2 traverses a greater linear distance than point 1
    (D) point 1 will make a complete revolution in less time than point 2
    (E) the product of the linear distance traversed by either point and the time required for one revolution is equal to the number of revolutions.

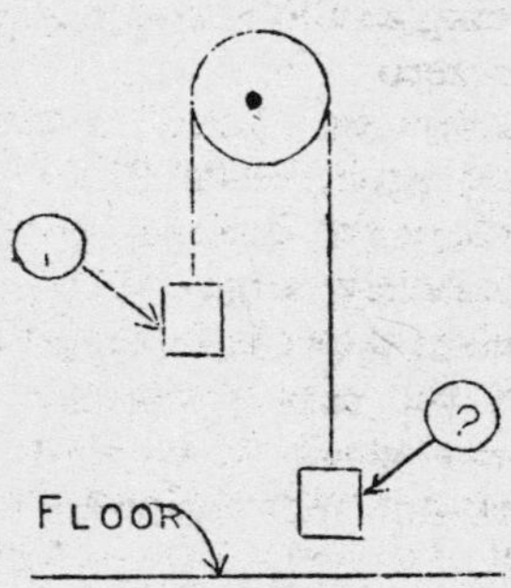

FIGURE V

34. Figure V represents a pulley, with practically no friction, from which two ten pound weights are suspended as indicated. If a downward force is applied to weight 1, it is most likely that weight 1 will
    (A) come to rest at the present level of weight 2
    (B) move downward until it is level with weight 2
    (C) move downward until it reaches the floor
    (D) pass weight 2 in its downward motion and then return to its present position
    (E) move downward a short distance before the direction of movement is reversed.

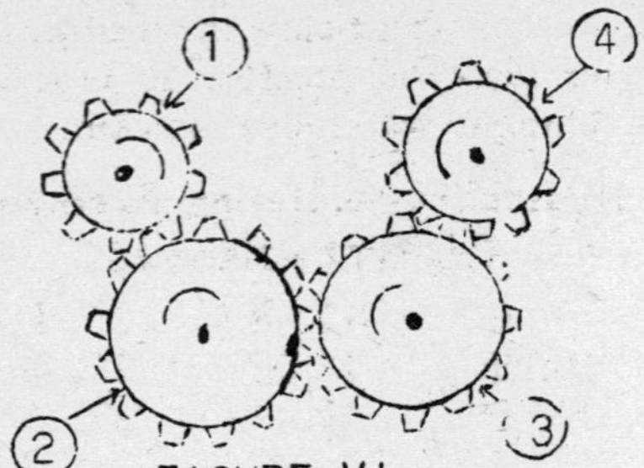

FIGURE VI

35. Figure VI represents four interlocking gear wheels. The wheels which turn in the same direction are
(A) 1 and 2
(B) 1 and 4
(C) 2 and 3
(D) 2 and 4
(E) 3 and 4.

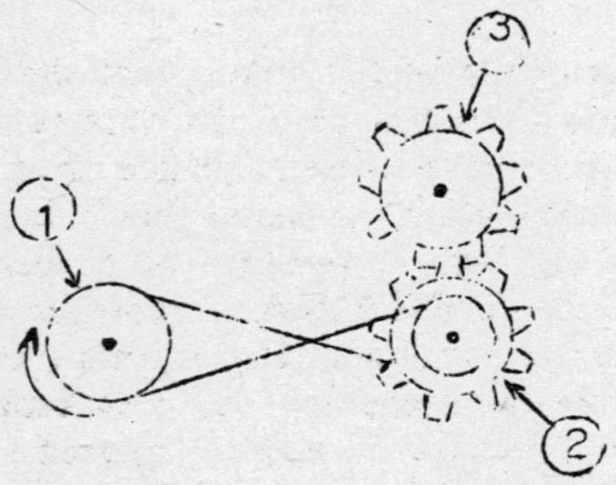

FIGURE VII

36. In Figure VII, the number 1 represents a wheel which is turning in a clockwise direction, indicated by the arrow. Wheel 1 is connected by a belt to wheel 2, which turns wheel 3. Of the following, the most accurate statement is that wheels
(A) 1 and 2 turn in the same direction
(B) 2 and 3 both turn in a clockwise direction
(C) 2 and 3 both turn in a direction opposite to clockwise
(D) 1 and 3 turn, but wheel 2 remains stationary
(E) 1 and 3 turn in the same direction.

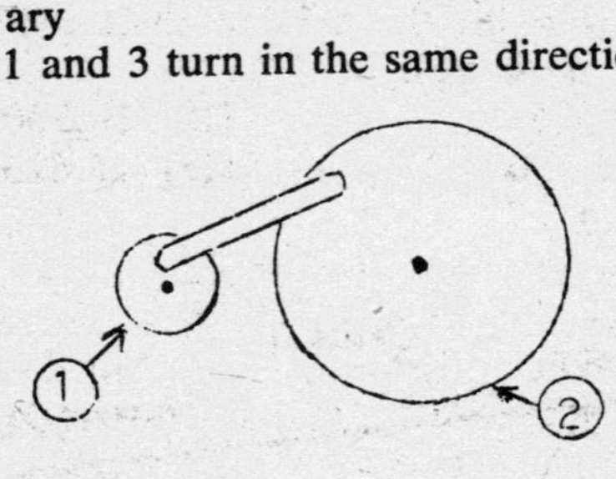

FIGURE VIII

37. Figure VIII represents two wheels, drawn to scale. Each wheel is free to revolve around its center, which is fixed. The two wheels are attached by a steel rod, as indicated. Wheel 2 has a diameter three times the diameter of wheel 1. Of the following, the most accurate statement is that, if wheel 1 makes one complete revolution, then wheel 2 will
(A) move back and forth
(B) make one complete revolution
(C) make three revolutions
(D) make one-third of a revolution in the opposite direction
(E) revolve at three times the speed of wheel 1.

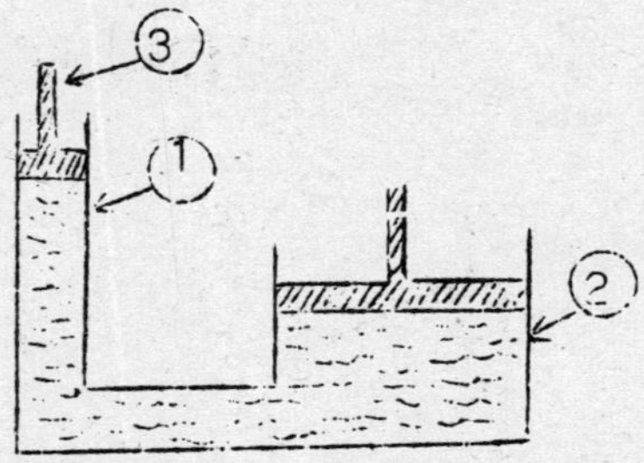

FIGURE IX

38. Figure IX represents two cylinders, 1 and 2, which contain water and are connected by a pipe. If the piston, 3, in cylinder 1 is lowered the
(A) piston in cylinder 2 will fall proportionately
(B) air in the connecting pipe will be forced to expand
(C) water in the system will occupy a greater volume
(D) density of water in the connecting pipe will be decreased proportionately
(E) level of the water in cylinder 2 will rise.

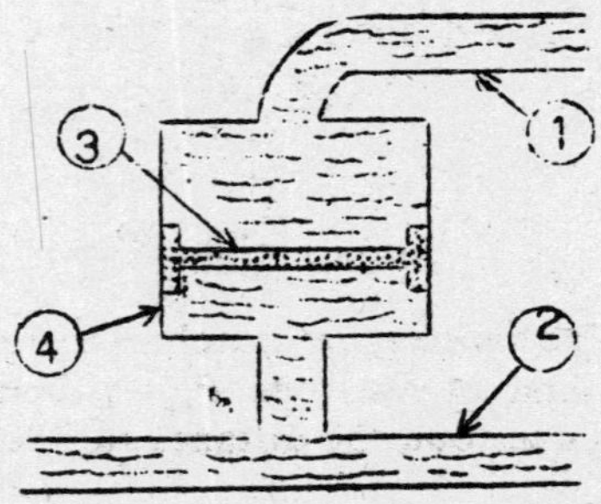

FIGURE X

39. In Figure X, the numbers 1 and 2 represent two pipes, both filled with water. A waterproof diaphragm, 3, slides up and down along the walls of a chamber, 4. Of the following, the most accurate statement is that if the water pressure in
    (A) pipe 1 exceeds the pressure in pipe 2, the diaphragm will move upwards
    (B) pipe 2 exceeds the pressure in pipe 1, the diaphragm will move downward
    (C) pipe 1 is equal to the pressure in pipe 2, the diaphragm will remain stationary
    (D) the chamber above the diaphragm is the same as the pressure in pipe 1, the diaphragm will remain stationary
    (E) the chamber below the diaphragm is greater than the pressure in pipe 2, the diaphragm will move upward.

Hydraulics—Entering Firemen

Correct Answers

1. A	9. C	17. B	25. D	33. C
2. B	10. B	18. C	26. E	34. C
3. A	11. D	19. A	27. A	35. D
4. C	12. A	20. C	28. B	36. E
5. D	13. E	21. E	29. E	37. A
6. B	14. E	22. D	30. C	38. E
7. E	15. B	23. B	31. B	39. C
8. D	16. D	24. E	32. D	

MECHANICAL INSIGHT FOR FIREMEN

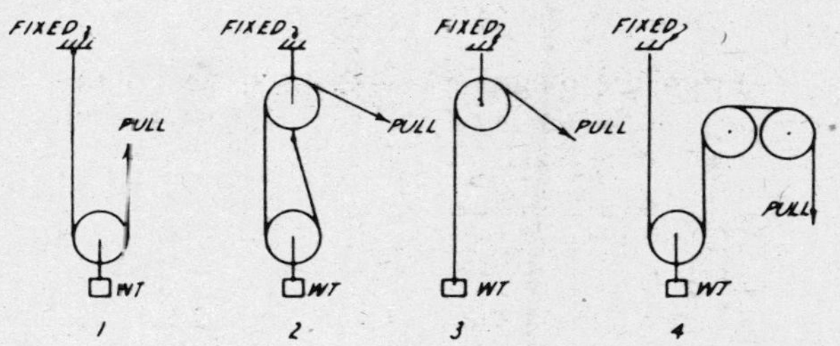

1. An object is to be lifted by means of a system of lines and pulleys. Of the systems shown above, the one which would require the *greatest* force to be used in lifting the weight is the one numbered
   (A) 1 (B) 2
   (C) 3 (D) 4.

2. In the accompanying sketch of a block and fall, if the end of the rope P is pulled so that it moves one foot, the distance the weight will be raised is

   (A) ½ ft. (B) 1 ft.
   (C) 1½ ft. (D) 2 ft.

3. A principle of lighting is that the intensity of illumination at a point is inversely proportional to the square of the distance from the source of illumination. Assume that a pulley lamp is lowered from a position of 6 feet to one of three feet above a desk. According to the above principle, we would expect that the amount of illumination reaching the desk from the lamp in the lower position, as compared to the higher position, will be
   (A) half as much
   (B) twice as much
   (C) four times as much
   (D) nine times as much.

4. Two parallel shafts with 27 inches between centers carry two stepped pulleys connected by a crossed belt. The smaller step on the driving shaft pulley is 5 inches in diameter. The driving shaft has a constant speed of 600 R.P.M. and the driven shaft has speeds of 150 R.P.M. and 1,000 R.P.M. The diameter of the larger step on the driving shaft pulley is, in inches, most nearly
   (A) 9.375 (B) 10.500
   (C) 12.750 (D) 15.625.

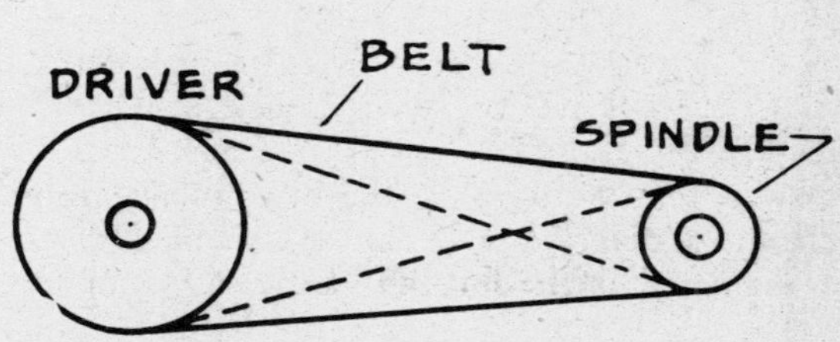

5. If the belt around the two pulleys is crossed as shown by the dotted lines
   (A) the speed of both the driver and spindle increase
   (B) both driver and spindle will change direction of motion
   (C) only the spindle will change direction of rotation
   (D) the driver will now be driven by the spindle.

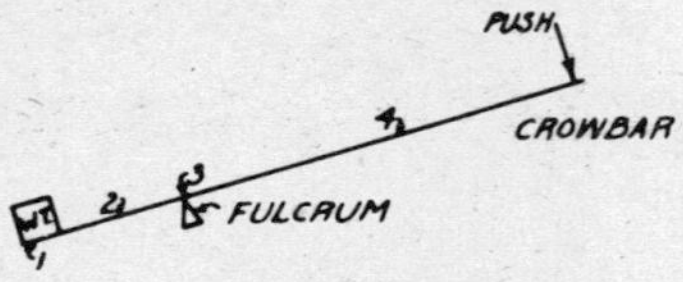

6. The above sketch shows a weight being lifted by means of a crowbar. The point at which the tendency for the bar to break is greatest is
   (A) 1 (B) 2
   (C) 3 (D) 4.

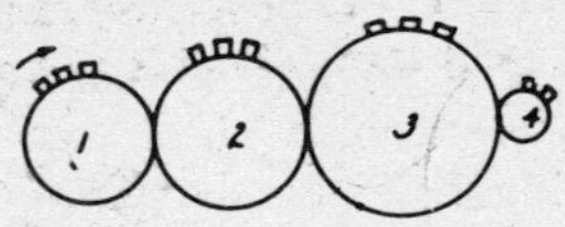

7. The accompanying sketch diagrammatically shows a system of meshing gears with relative diameters as drawn. If gear 1 is made to rotate in the direction of the arrow, then the gear that will turn fastest is numbered
   (A) 1 (B) 2
   (C) 3 (D) 4.

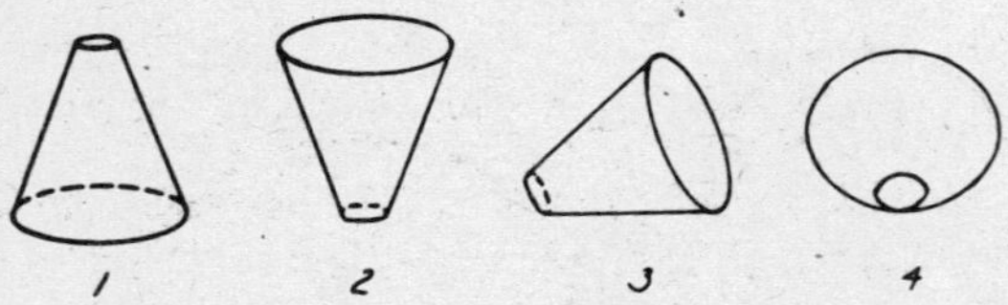

8. The above diagrams show four positions of the same object. The position in which this object is most stable is
   (A) 1 (B) 2
   (C) 3 (D) 4.

Answer items 9 to 13 on the basis of Diagram I. Schematic diagram I shows a "u" tube of uniform diameter open at both ends and containing three different liquids which will not mix: A, B and C. The numbers show cubic contents measured from the zero point.

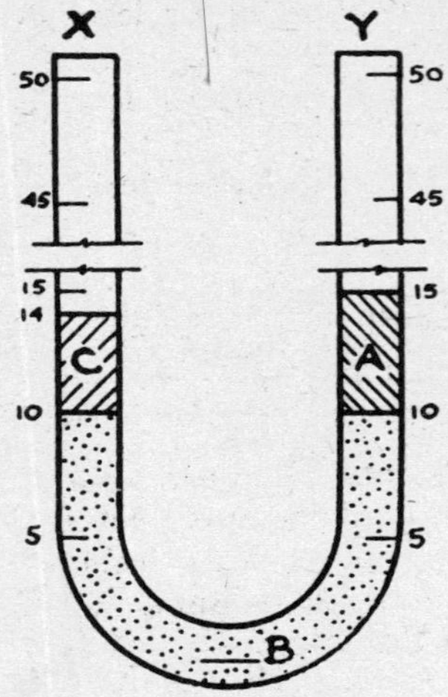

DIAGRAM I

9. From the diagram, it can be expected that in regard to density (weight per unit of volume)
   (A) A is equal to C
   (B) C is greater than B
   (C) A is greater than B
   (D) C is less than A
   (E) A is less than C.

10. If half of liquid A is removed from the tube,
   (A) liquid C will move down in tube X until its top level is at 11.5
   (B) liquid B will move up in tube Y until its top level is at 15
   (C) liquid C will move down in tube X an amount which is not determinable from the information given
   (D) liquid B will move up in tube Y until its top level is at 12.5
   (E) liquid C will move down in tube X until its top level is at 12.75.

11. If half of liquid B is removed from the tube by means of an outlet at the bottom, one result will be that
   (A) liquid A will move down until its top level is at 5
   (B) liquid C will move down until its top level is at 10
   (C) liquid C will move down an amount which is not determinable from the information given
   (D) liquid C will move down until its top level is at 5
   (E) liquid A will move down until its top level is at 10.

12. If half of liquid A and half of liquid C are removed at one time from the tube, liquid B will
   (A) remain in the same position
   (B) move until its upper level is at 8 in tube X
   (C) move until its upper level is at 12.25 in tube X
   (D) move until its upper level is at 7.75 in tube Y
   (E) move until its upper level is at 12.50 in tube Y.

13. If a quantity of liquid B equal to one-half the quantity already in the tube is poured into open end X, liquid A will
   (A) remain stationary
   (B) move until its upper level is at 17.5
   (C) move until its upper level is at 20
   (D) move until its upper level is at 25
   (E) move up an amount not determinable from the information given.

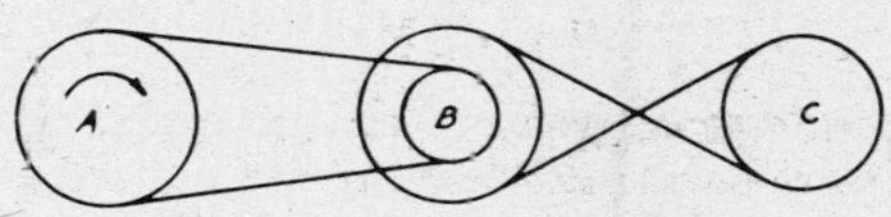

14. The above sketch diagrammatically shows a pulley and belt system. If pulley A is made to rotate in a clockwise direction, then pulley C will rotate
   (A) faster than pulley A and in a clockwise direction
   (B) slower than pulley A and in a clockwise direction
   (C) faster than pulley A and in a counter-clockwise direction
   (D) slower than pulley A and in a counter-clockwise direction.

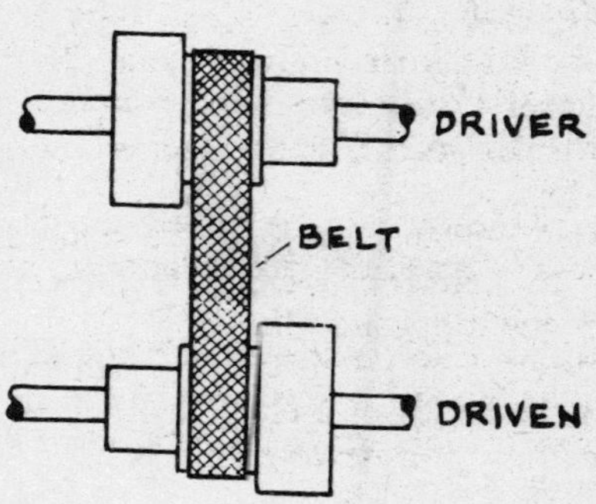

15. The pulleys are stepped in size to
   (A) vary the speed of the driven equipment
   (B) keep the shafts properly spaced
   (C) keep the belt from slipping off
   (D) take up belt slackness.

16. The leverage system in the sketch above is used to raise a weight. In order to reduce the amount of force required to raise the weight, it is necessary to
   (A) decrease the length of the lever
   (B) place the weight closer to the fulcrum
   (C) move the weight closer to the person applying the force
   (D) move the fulcrum further from the weight.

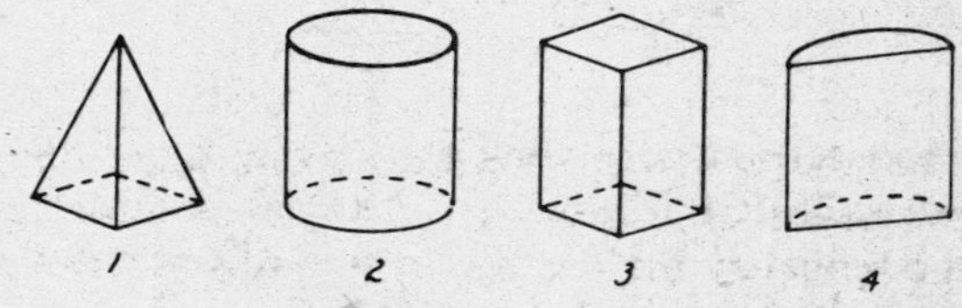

17. The above sketches show four objects which weigh the same but have different shapes. The object which is most difficult to tip over is numbered
   (A) 1 (B) 2
   (C) 3 (D) 4.

Answer items 18 to 22 on the basis of Diagram II.

A, B, C and D are four meshed gears forming a gear train. Gear A is the driver. Gears A and D each have twice as many teeth as gear B, and gear C has four times as many teeth as gear B. The diagram is schematic; the teeth go all around each gear.

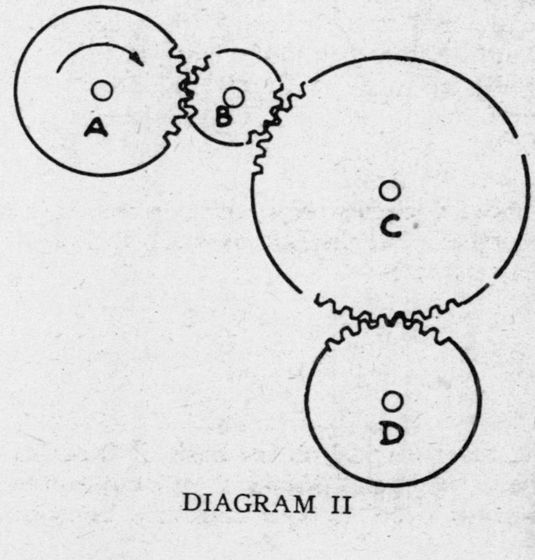

DIAGRAM II

18. Two gears which turn in the same direction are
(A) A and B (B) B and C
(C) C and D (D) D and A
(E) B and D.

19. The two gears which revolve at the same speed are gears
(A) A and C (B) A and D
(C) B and C (D) B and D
(E) D and C.

20. If all the teeth on gear C are stripped without affecting the teeth on gears A, B, and D, then rotation would occur only in
(A) gear C (B) gear D
(C) gears A and B (D) gears A, B and D
(E) gears B and D.

21. If gear D is rotating at the rate of 100 R.P.M., then gear B is rotating at the rate of
(A) 25 R.P.M. (B) 50 R.P.M.
(C) 100 R.P.M. (D) 200 R.P.M.
(E) 400 R.P.M.

22. If gear A turns at the rate of two revolutions per second, then the number of revolutions per second that gear C turns is
(A) 1 (B) 2 (C) 3 (D) 4 (E) 8.

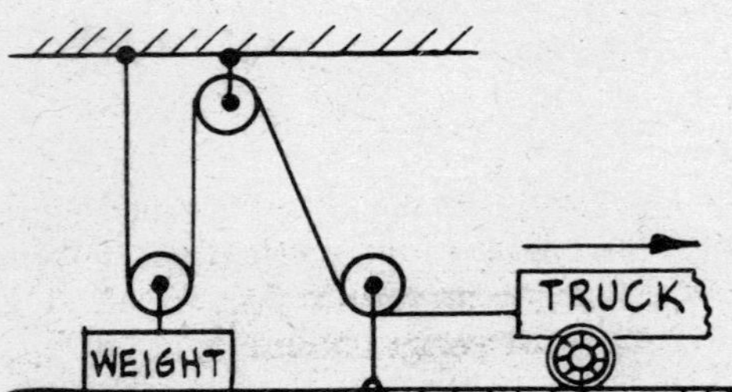

23. The weight is to be raised by means of the rope attached to the truck. If the truck moves forward 20 feet, then the weight will rise
(A) 20 feet (B) 15 feet
(C) 10 feet (D) 5 feet.

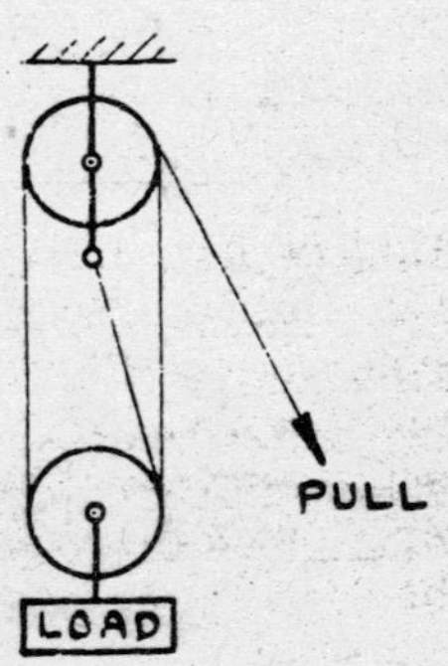

24. In raising a given load with the block and fall shown, the required pull decreases
(A) if larger pulley blocks are used
(B) as the distance between the pulley blocks decreases
(C) as the number of pulleys in each block is increased
(D) if the size of the rope is increased.

Figure I represents a schematic drawing of part of the operating mechanism of a piece of Fire Department equipment. Wheels 1 and 2 have the same diameter. Wheel 3 has a diameter one-third the diameter of wheel 1. Wheel 1 is connected to wheel 3 by a belt and wheel 2 is rigidly attached to wheel 3, as illustrated. Wheel 1 turns in the same direction as the hands of a clock, at a speed of 10 revolutions per minute.

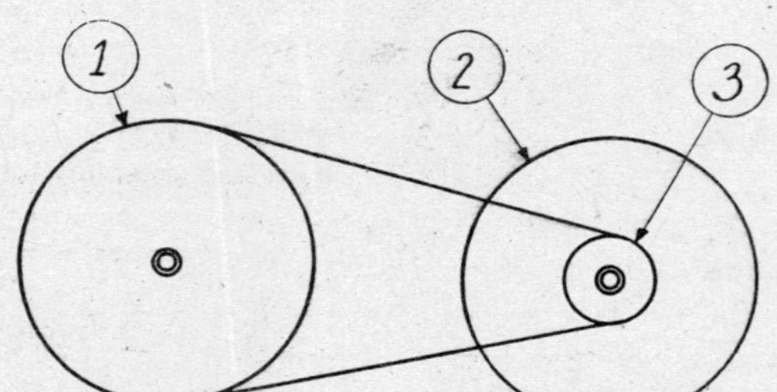

Figure I

Answer questions 25 thru 27 on the basis of this drawing.

25. Since wheel 1 turns in the direction of the hands of a clock, then
(A) wheel 3 turns in a direction opposite to wheel 1
(B) wheel 2 turns in the same direction as wheel 1
(C) wheel 3 turns in a direction opposite to wheel 2
(D) wheels 1 and 2 turn in the same direction as the hands of a clock, but wheel 3 turns in the opposite direction
(E) wheels 1 and 3 turn in the same direction as the hands of a clock, but wheel 2 turns in the opposite direction.

26. Since wheel 1 turns at a rate of 10 revolutions per minute, the number of revolutions per minute made by wheel 3 is, most nearly,
(A) 3 (B) 10
(C) 30 (D) 60
(E) 90.

27. If the mechanism in Figure I is changed so that wheel 1 is connected directly to wheel 2 by gear teeth instead of to wheel 3, as illustrated, then the most accurate statement is that
    (A) the direction of turning of wheel 2 will remain unchanged
    (B) the speed of turning of wheel 2 will remain unchanged
    (C) the direction of turning of wheel 3 will be changed, but the speed of turning will remain unchanged
    (D) the speed of turning of wheel 3 will be changed, but the direction of turning will remain unchanged
    (E) both the speed and direction of turning of wheel 3 will be changed.

The device shown below represents schematically a mechanism commonly used to change reciprocating (back and forth) motion to rotating (circular) motion.

The following questions, numbered 28 to 30 inclusive, are to be answered with reference to this device.

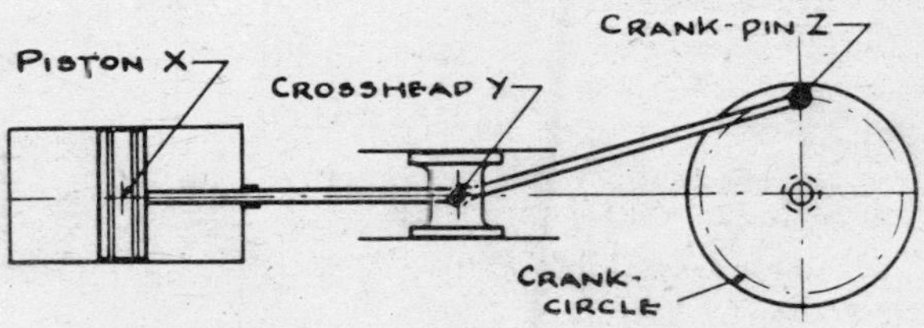

28. Assume that piston X is placed in its extreme left position so that X, Y and Z are in a horizontal line. If a horizontal force to the right is applied to the piston X, we may then expect that
    (A) the crank-pin Z will revolve clockwise
    (B) the crosshead Y will move in a direction opposite to that of X
    (C) the crank-pin Z will revolve counterclockwise
    (D) no movement will take place
    (E) the crank-pin Z will oscillate back and forth.

29. If we start from the position shown in the above diagram and move piston X to the right, the result will be that
    (A) the crank-pin Z will revolve counterclockwise and crosshead Y will move to the left
    (B) the crank-pin Z will revolve clockwise and crosshead Y will move to the left
    (C) the crank-pin Z will revolve counterclockwise and crosshead Y will move to the right
    (D) the crank-pin Z will revolve clockwise and crosshead Y will move to the right
    (E) crosshead Y will move to the left as piston X moves to the right.

30. If crank-pin Z is moved closer to the center of the crank circle, then the length of the
    (A) stroke of piston X is increased
    (B) stroke of piston X is decreased
    (C) stroke of piston X is unchanged
    (D) rod between the piston X and crosshead Y is increased
    (E) rod between the piston X and crosshead Y is decreased.

The figure below represents schematically a block and fall tackle. The advantage derived from this machine is that the effect of the applied force is multiplied by the number of lines of rope directly supporting the load. The following two questions, numbered 31 and 32, are to be answered with reference to this figure.

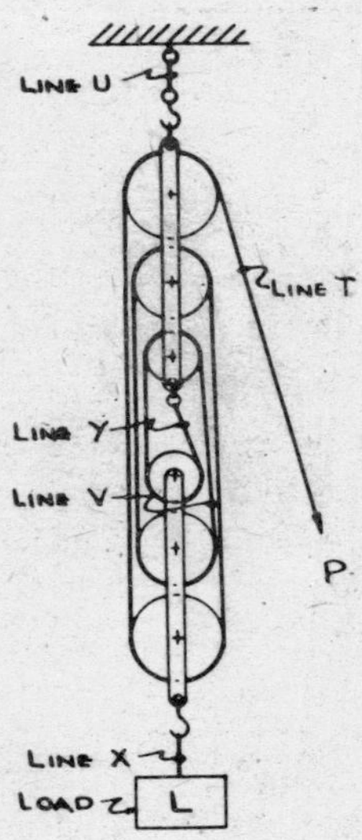

31. Pull P is exerted on line T to raise the load L. The line in which the largest strain is finally induced is line
    (A) T
    (B) U
    (C) V
    (D) X
    (E) Y

32. If the largest pull P that two men can apply to line T is 280 lbs., the maximum load L that they can raise without regard to frictional losses is most nearly
(A) 1960 lbs. (B) 1680 lbs.
(C) 1400 lbs. (D) 1260 lbs.
(E) 1120 lbs.

33. As a demonstration, firemen set up two hose lines identical in every respect except that one was longer than the other. Water was then delivered through these lines from one pump and it was seen that the stream from the longer hose line had a shorter "throw." Of the following, the most valid explanation of this difference in "throw" is that the
(A) air resistance to the water stream is proportional to the length of hose
(B) time required for water to travel through the longer hose is greater than for the shorter one
(C) loss due to friction is greater in the longer hose than in the shorter one
(D) rise of temperature is greater in the longer hose than in the shorter one
(E) longer hose line probably developed a leak at one of the coupling joints.

Mechanical Insight

Correct Answers

(You'll learn more by writing your own answers before comparing them with these.)

1. C	10. C	19. B	28. D
2. A	11. E	20. C	29. D
3. C	12. A	21. D	30. B
4. D	13. C	22. A	31. B
5. C	14. C	23. C	32. B
6. C	15. A	24. C	33. C
7. D	16. B	25. B	
8. A	17. A	26. C	
9. E	18. E	27. E	

REASONING AND JUDGMENT QUESTIONS FOR FIREMEN

1. Of the following, the chief advantage of having firemen under competitive civil service is that
   (A) fewer fires tend to occur
   (B) fire prevention becomes a reality instead of a distant goal
   (C) the efficiency level of the personnel tends to be raised
   (D) provision may then be made for training by the municipality
   (E) responsibility is necessarily centralized.

2. Of the following, it is least likely that fire will be caused by
   (A) arson
   (B) poor building construction
   (C) carelessness
   (D) inadequate supply of water
   (E) explosives.

3, 4, 5. The following three questions, numbered 3 through 5 inclusive, are to be answered on the basis of the table below. Data for certain categories have been omitted from the table. You are to calculate the missing numbers if needed to answer the questions.

	Last Year	This Year	Increase
Firemen	9326	9744	
Lieutenants	1355	1417	
Captains		443	107
Others			
TOTAL	11469	12099	

3. The number in the "Others" group last year was most nearly
   (A) 450 (B) 475 (C) 500 (D) 525

4. The group which had the largest percentage was
   (A) Firemen (B) Lieutenants
   (C) Captains (D) Others.

5. This year the ratio between Firefighters and all other ranks of the uniformed force was most nearly
   (A) 5:1 (B) 4:1
   (C) 2:1 (D) 1:1.

6. "It is said that in big cities a fire alarm box can be seen from any corner." This is a desirable condition mainly because
   (A) several alarms may be sounded by one person running from one box to another
   (B) little time is lost in sounding an alarm
   (C) an alarm may be sounded from a different box if the nearest one is out of order
   (D) fire apparatus can quickly reach any box
   (E) the number of fires is reduced.

7. The primary function of a fire department is to
   (A) maintain peace
   (B) reduce industrial conflict
   (C) improve the governmental machinery
   (D) preserve life and property
   (E) promote good citizenship.

8. The problem of determining the relative importance of brain and brawn in the equipment of a fireman may best be settled by
   (A) quantitative investigation
   (B) securing the opinions of prominent educators
   (C) ascertaining the attitude of the average citizen
   (D) flipping a coin or by lot
   (E) averaging both so that neither has any weight.

9. Suppose that you are employed as a fireman. A new man is appointed to your company. This man has red hair. It is most probable that this man will
   (A) have a fiery, uncontrollable temper
   (B) be very aggressive
   (C) be very intelligent
   (D) be much like the other men in your company
   (E) be unusually obstinate.

10. Suppose hydrants with a flowing capacity of less than 500 gals. per minute were to be painted red, hydrants with a flowing capacity between 500 and 1000 gallons were to be painted yellow, and hydrants with a flowing capacity of 1000 gallons or greater per minute were to be painted green. The principal advantage of such a scheme is that
(A) fewer fires would occur
(B) more water would become available at a fire
(C) citizens would become more acutely aware of the importance of hydrants
(D) parking in front of hydrants would be reduced
(E) firemen would save time.

11. "At two o'clock in the morning Mrs. Smart wakened her husband and said the house was on fire. Mr. Smart dressed hurriedly, ran seventeen blocks past five fire alarm boxes to the fire station, and told the firemen that his house was on fire. When Mr. Smart and the firemen had returned, the house had burned down." This is an illustration of the
(A) need for a plentiful supply of fire alarm boxes
(B) fact that fire is no respecter of persons
(C) necessity of preventing fires
(D) desirability of educating the public
(E) need for more fire stations.

12. A fire breaks out simultaneously in six different parts of a large building. It is most reasonable to believe that the fire is a result of
(A) arson (B) carelessness
(C) spontaneous combustion (D) explosives
(E) enzymic action.

13. Suppose that the total number of fires reported during May is equal to F and that the number of fires reported by telephone during that time is P. The ratio of the number of fires reported by telephone to the total number of fires reported is R. Then the product of R and F is equal to the
(A) ratio of fires not reported by telephone
(B) number of fires not reported by telephone
(C) number of fires reported by telephone
(D) ratio of fires reported by telephone
(E) total number of fires reported.

14. "The fireman must not only know how to make the proper knots, but he must know how to make them quickly." Of the following, the chief justification for this statement is that
(A) a fireman uses many kinds of knots
(B) a slowly tied knot slips more readily than one tied quickly
(C) haste makes waste
(D) it is more important for a knot to be tied correctly than quickly
(E) in most fire operations, speed is important.

15. A survey of dip tank fires in large cleaning establishments shows that the majority of these fires started with such an obvious hazard as cutting and welding torches. This experience indicates most strongly a need for
(A) a law prohibiting the use of welding torches
(B) fire-safety programs in this industry
(C) a law requiring welders to be licensed
(D) eliminating the use of dip tanks
(E) the location of additional fire companies in the neighborhood of dip tank users.

16. In order to reduce the number of collisions between fire apparatus and private automobiles, a number of suggestions have been made regarding the training and selection of drivers. Of the following courses of action, the one which is *least* likely to lead to the reduction of accidents is to
(A) select drivers who have the longest driving experience
(B) require all members of a company to drive fire apparatus
(C) select drivers who have experience in driving trucks
(D) select an alternate driver for every piece of fire apparatus
(E) require drivers to practice driving and maneuvering fire apparatus.

17. "Automobile parking on city streets is becoming daily a greater menace to effective fire fighting." Of the following effects of automobile parking, the *least* serious is that
(A) access to the fire building may be obstructed

(B) ladder and rescue work may be retarded
(C) traffic congestion may make it difficult for fire apparatus to get through
(D) greater lengths of hose may be needed in fighting a fire
(E) street width is reduced so that fire apparatus has difficulty in maneuvering.

18. "Sometimes at a fire a man endures great punishment in order to reach a desired position which could have been reached more quickly and with less hardship." Of the following, the chief implication of the above statement is that
(A) one should take time off during a fire to think completely through problems presented
(B) courage is an important asset in a fireman
(C) training is an important factor in firefighting
(D) modern firefighting requires alert firefighters
(E) no two fires are ever exactly the same.

19. The statement "Municipal inspections should be coordinated" suggests most nearly that
(A) fire hazards are dangerous
(B) the detection of fire hazards is connected with other problems to be found in a municipality
(C) municipal and fire inspection services are of unequal importance
(D) municipal inspections are closely tied up with non-inspectional services
(E) municipal inspectors should cooperate with members of the community.

20. "With fireproof schools, it would appear that drills are unnecessary." The main reason for believing this statement to be false is that
(A) panics sometime occur
(B) fire extinguishers are available in every school
(C) fire alarms are easily sounded
(D) children are accustomed to drilling
(E) children should be taught that fireproofing is desirable.

21. "Flame is of varying heat according to the nature of the substance producing it" means most nearly that
(A) some fires are larger than others
(B) the best measure of the heat produced by a particular substance is its temperature
(C) there can be no fire without flame or flame without fire
(D) inflammable substances should not be used in building construction
(E) the degree of heat evolved by the combustion of different materials is not identical.

22. "Fire records indicate that in many fires caused by heating equipment, insufficient clearance from combustible materials is an important factor." From this observation it follows that the major reason, of those listed, that fires increase in cold weather is that
(A) cold weather makes combustible materials more inflammable
(B) combustible materials tend to expand when heated
(C) heating equipment is in operation
(D) hot air from heating equipment is more inflammable
(E) heating equipment tends to become defective.

23. "A fireman does not have time to read engineering books when human life is in jeopardy" means most nearly that
(A) preparation for an emergency cannot be made at the particular instant when the emergency arises
(B) the technique of fire fighting bears no relation to engineering findings
(C) life is more important than property
(D) firemen do not have enough time to read books
(E) reading is a useless pastime for a fireman.

24. "A fire will occur when a flammable material is sufficiently heated in the presence of oxygen." It follows from this statement that a fire will occur whenever there is sufficient
(A) heat, hydrogen and fuel
(B) oxygen, hydrogen and fuel
(C) heat, air and fuel
(D) gasoline, heat and fuel
(E) oxygen, air and heat.

25. Suppose that the vast majority of the fires which had taken place in 1940 in New York City occurred in a particular type of building structure. Before a fireman could reasonably infer that this particular type of building structure was unduly susceptible to fires in New York City in 1940, he would have to know
   (A) the precise incidence, in terms of per cent, that is represented by the somewhat loose phrase "vast majority"
   (B) the frequency of this type of building structure as compared with that of other types of building structures
   (C) whether 1940 is validly to be taken as the fiscal year or the calendar year
   (D) whether New York City may legitimately be considered as representative of the country as a whole in regard to the statistical frequency of this type of building structure
   (E) whether 1940 is non-representative in any respect as a computational base.

26. "Every fire is a potential conflagration." Of the following, the most valid inference that may be drawn from this statement is that
   (A) no matter how insignificant a fire may appear to be, a first effort should be to isolate it
   (B) the method of fighting a potential fire must be adapted to the unique circumstances surrounding that fire
   (C) the apparatus and men sent immediately to a fire should be sufficient to handle practically any conflagration that may be found
   (D) the full potentialities of a fire are usually realized
   (E) none of the foregoing is true.

27. One of the firemen in your company is far superior to all the others in ability to learn new material quickly. It follows that
   (A) he will be below average in physical strength and muscular coordination
   (B) he will forget the material he learns more quickly than those who learn slowly but surely
   (C) he is an avid reader
   (D) he is considerably younger than the other firemen
   (E) none of the foregoing are necessarily true.

28. Suppose that, for a certain period of time studied, the percentage of telephone alarms which were false alarms was less than the percentage of fire box alarms which were false alarms. Then the one of the following which is most accurate, solely on the basis of the above statement, is that during the period studied
   (A) more alarms were transmitted by telephone than by fire box
   (B) more alarms were transmitted by fire box than by telephone
   (C) relatively fewer false alarms were transmitted by telephone than by fire box
   (D) relatively fewer false alarms were transmitted by fire box than by telephone
   (E) the number of false telephone alarms was less than the number of false fire box alarms.

29. "There are more engine companies than hook and ladder companies. However, to conclude that the number of firemen assigned to engine companies exceeds the number of firemen assigned to hook and ladder companies is to make a basic assumption." Of the following, the most accurate statement of the basic assumption referred to in this quotation is that
   (A) the number of hook and ladder companies does not differ greatly from the number of engine companies
   (B) about the same number of men, on the average, are assigned to each type of company
   (C) an engine company, on the average, has fewer men than a hook and ladder company
   (D) the largest engine company is no larger than the largest hook and ladder company
   (E) there is no fixed relationship between the size of engine companies and hook and ladder companies.

30. If the fire company due to arrive first at a fire arrived one minute before the company second due, and the company third due arrived four minutes after the alarm was sent, then the one of the following statements which is most accurate is that the
   (A) second due company arrived three minutes before the third due company
   (B) third due company arrived three minutes after the first due company
   (C) second due company arrived two minutes after the alarm was sent

(D) second due company arrived one minute after the first due company
(E) first due company arrived five minutes before the third due company.

31. "During June 25 per cent of all the fires in a Western city were in buildings of type A, 40 per cent were in buildings of type B, and 15 per cent were in buildings of type C." Of the following, the most accurate statement is that the total number of fires during June was
(A) equal to the sum of the percentages of fires in the three types of buildings, divided by 100
(B) less than 100 per cent
(C) one-fourth the number of fires in buildings of type B
(D) equal to the sum of the number of fires in buildings of types A, B, and C
(E) four times the number of fires in buildings of type A.

32. Suppose that the number of fires during May increased 5 per cent over the number of fires during April, but that the number of fires during June decreased 5 per cent as compared with the number of fires during May. Then the one of the following statements which is most accurate is that there were
(A) exactly as many fires during April as there were during June
(B) more fires during June than during May
(C) more fires during April than during May
(D) fewer fires during June than during April
(E) fewer fires during May than during June.

33. Suppose that two 60-foot ladders have been placed against the side of a building. The base of the first ladder is 10 feet from the building and the base of the second ladder is 20 feet from the building. Then, of the following, the most accurate statement is that
(A) the first ladder will make a larger angle with the ground than the second ladder will make
(B) the top of the second ladder is higher above the ground than the top of the first ladder
(C) both ladders will reach to the same height above the ground
(D) the first ladder will make a larger angle with the building than the second ladder will make
(E) both ladders will make the same angle with the building.

34. Suppose that water is being pumped from an engine into a hose at a pressure of R pounds and that the friction loss in the hose is P pounds per foot of hose. The pressure loss in the hose due to friction will equal the pump pressure when the length of the hose is equal to
(A) P divided by R feet (B) R minus P feet
(C) 12 times P feet (D) R feet
(E) R divided by P feet.

35. It is desirable that the fire department have ladders of varying lengths mainly because some
(A) fires occur at greater distances from the ground than others
(B) firemen are more agile than others
(C) firemen are taller than others
(D) fires are harder to extinguish than others
(E) buildings are not fireproof.

36. The lever on fire alarm boxes for use by citizens should be
(A) very, very easy to manipulate
(B) just a little difficult to manipulate
(C) very difficult to manipulate
(D) constructed without regard to ease of manipulation
(E) constructed with regard only to ease of manipulation.

Correct Answers

Reasoning and Judgment

1. C	10. E	19. B	28. C
2. D	11. D	20. A	29. B
3. A	12. A	21. E	30. D
4. C	13. C	22. C	31. E
5. B	14. E	23. A	32. D
6. B	15. B	24. C	33. A
7. D	16. B	25. B	34. E
8. A	17. D	26. A	35. A
9. D	18. C	27. E	36. B

PART 4

4

Practice and Prepararation for the General Test Subjects of the Examination

The past examinations conducted for firefighter contained sections on arithmethic, vocabulary, English usage and general testing. Recently many jurisdictions have dropped these subjects. However, the sections for preparation in these areas are still included in this book for two reasons: your jurisdiction may still have these subjects included in its examination and, if not, using this part of the book will provide the mental gymnastics necessary to sharpen your mind for the actual examination. It will also help you understand the necessity of being careful and patient when taking examinations and will enhance your opportunities for promotion once you have been appointed to the position you are being tested for.

SAMPLE ANSWER SHEET

EXAM TITLE	TODAY'S DATE	
SCHOOL OR BUILDING	ROOM NO.	SEAT NO.

PRINT WITH SOFT PENCIL ONLY. Print, with pencil, your SOCIAL SECURITY NO. and the EXAM. NO. in the boxes at the tops of the columns. ONE NUMBER IN A BOX. In each column, darken (with pencil) the oval containing the number in the box at the top of the column. Only ONE OVAL in a COLUMN should be darkened. Then, using your pencil, print in the: Exam. Title, School or Building, Room No., Seat No., and Today's Date.

Follow the instructions given in the question booklet. Mark nothing ▶ but your answers in the ovals below.
SAMPLE QUESTION: When we add 5 and 3 we get: (A) 11 (B) 9 (C) 8 (D) 2. Since the answer is 8, your answer should be marked like this: (A) (B) ● (D)
WARNING: Be sure that the oval you fill in is in the row numbered the same as the question you are answering. Use a No. 2 pencil (soft pencil).
BE SURE YOUR PENCIL MARKS ARE HEAVY AND BLACK. ERASE COMPLETELY ANY ANSWER YOU WISH TO CHANGE. DO NOT make stray pencil dots, dashes or marks ANYPLACE on this SHEET.

YOUR SOCIAL SECURITY NO.								
0	0	0	0	0	0	0	0	0
1	1	1	1	1	1	1	1	1
2	2	2	2	2	2	2	2	2
3	3	3	3	3	3	3	3	3
4	4	4	4	4	4	4	4	4
5	5	5	5	5	5	5	5	5
6	6	6	6	6	6	6	6	6
7	7	7	7	7	7	7	7	7
8	8	8	8	8	8	8	8	8
9	9	9	9	9	9	9	9	9

EXAM NO.			
0	0	0	0
1	1	1	1
2	2	2	2
3	3	3	3
4	4	4	4
5	5	5	5
6	6	6	6
7	7	7	7
8	8	8	8
9	9	9	9

START HERE

1 (A) (B) (C) (D)	26 (A) (B) (C) (D)	51 (A) (B) (C) (D)	76 (A) (B) (C) (D)	101 (A) (B) (C) (D)	126 (A) (B) (C) (D)
2 (A) (B) (C) (D)	27 (A) (B) (C) (D)	52 (A) (B) (C) (D)	77 (A) (B) (C) (D)	102 (A) (B) (C) (D)	127 (A) (B) (C) (D)
3 (A) (B) (C) (D)	28 (A) (B) (C) (D)	53 (A) (B) (C) (D)	78 (A) (B) (C) (D)	103 (A) (B) (C) (D)	128 (A) (B) (C) (D)
4 (A) (B) (C) (D)	29 (A) (B) (C) (D)	54 (A) (B) (C) (D)	79 (A) (B) (C) (D)	104 (A) (B) (C) (D)	129 (A) (B) (C) (D)
5 (A) (B) (C) (D)	30 (A) (B) (C) (D)	55 (A) (B) (C) (D)	80 (A) (B) (C) (D)	105 (A) (B) (C) (D)	130 (A) (B) (C) (D)
6 (A) (B) (C) (D)	31 (A) (B) (C) (D)	56 (A) (B) (C) (D)	81 (A) (B) (C) (D)	106 (A) (B) (C) (D)	131 (A) (B) (C) (D)
7 (A) (B) (C) (D)	32 (A) (B) (C) (D)	57 (A) (B) (C) (D)	82 (A) (B) (C) (D)	107 (A) (B) (C) (D)	132 (A) (B) (C) (D)
8 (A) (B) (C) (D)	33 (A) (B) (C) (D)	58 (A) (B) (C) (D)	83 (A) (B) (C) (D)	108 (A) (B) (C) (D)	133 (A) (B) (C) (D)
9 (A) (B) (C) (D)	34 (A) (B) (C) (D)	59 (A) (B) (C) (D)	84 (A) (B) (C) (D)	109 (A) (B) (C) (D)	134 (A) (B) (C) (D)
10 (A) (B) (C) (D)	35 (A) (B) (C) (D)	60 (A) (B) (C) (D)	85 (A) (B) (C) (D)	110 (A) (B) (C) (D)	135 (A) (B) (C) (D)
11 (A) (B) (C) (D)	36 (A) (B) (C) (D)	61 (A) (B) (C) (D)	86 (A) (B) (C) (D)	111 (A) (B) (C) (D)	136 (A) (B) (C) (D)
12 (A) (B) (C) (D)	37 (A) (B) (C) (D)	62 (A) (B) (C) (D)	87 (A) (B) (C) (D)	112 (A) (B) (C) (D)	137 (A) (B) (C) (D)
13 (A) (B) (C) (D)	38 (A) (B) (C) (D)	63 (A) (B) (C) (D)	88 (A) (B) (C) (D)	113 (A) (B) (C) (D)	138 (A) (B) (C) (D)
14 (A) (B) (C) (D)	39 (A) (B) (C) (D)	64 (A) (B) (C) (D)	89 (A) (B) (C) (D)	114 (A) (B) (C) (D)	139 (A) (B) (C) (D)
15 (A) (B) (C) (D)	40 (A) (B) (C) (D)	65 (A) (B) (C) (D)	90 (A) (B) (C) (D)	115 (A) (B) (C) (D)	140 (A) (B) (C) (D)
16 (A) (B) (C) (D)	41 (A) (B) (C) (D)	66 (A) (B) (C) (D)	91 (A) (B) (C) (D)	116 (A) (B) (C) (D)	141 (A) (B) (C) (D)
17 (A) (B) (C) (D)	42 (A) (B) (C) (D)	67 (A) (B) (C) (D)	92 (A) (B) (C) (D)	117 (A) (B) (C) (D)	142 (A) (B) (C) (D)
18 (A) (B) (C) (D)	43 (A) (B) (C) (D)	68 (A) (B) (C) (D)	93 (A) (B) (C) (D)	118 (A) (B) (C) (D)	143 (A) (B) (C) (D)
19 (A) (B) (C) (D)	44 (A) (B) (C) (D)	69 (A) (B) (C) (D)	94 (A) (B) (C) (D)	119 (A) (B) (C) (D)	144 (A) (B) (C) (D)
20 (A) (B) (C) (D)	45 (A) (B) (C) (D)	70 (A) (B) (C) (D)	95 (A) (B) (C) (D)	120 (A) (B) (C) (D)	145 (A) (B) (C) (D)
21 (A) (B) (C) (D)	46 (A) (B) (C) (D)	71 (A) (B) (C) (D)	96 (A) (B) (C) (D)	121 (A) (B) (C) (D)	146 (A) (B) (C) (D)
22 (A) (B) (C) (D)	47 (A) (B) (C) (D)	72 (A) (B) (C) (D)	97 (A) (B) (C) (D)	122 (A) (B) (C) (D)	147 (A) (B) (C) (D)
23 (A) (B) (C) (D)	48 (A) (B) (C) (D)	73 (A) (B) (C) (D)	98 (A) (B) (C) (D)	123 (A) (B) (C) (D)	148 (A) (B) (C) (D)
24 (A) (B) (C) (D)	49 (A) (B) (C) (D)	74 (A) (B) (C) (D)	99 (A) (B) (C) (D)	124 (A) (B) (C) (D)	149 (A) (B) (C) (D)
25 (A) (B) (C) (D)	50 (A) (B) (C) (D)	75 (A) (B) (C) (D)	100 (A) (B) (C) (D)	125 (A) (B) (C) (D)	150 (A) (B) (C) (D)

FIREFIGHTER, F. D.

TOP SCORES ON READING TESTS

Students must be able to read the paragraphs quickly, and still be able to answer questions correctly. The more correct answers you can give, the better your score will be. But if there are twenty paragraphs, and you are able to finish only ten because you read slowly, obviously, you are going to get a score of 50 percent, even if you answer all the questions correctly. On the other hand, if you finish all the paragraphs but can only answer half of the questions correctly, you will still get only 50 percent. Your goal, then, is to build up enough speed to finish all the paragraphs, and at the same time give as many correct answers as possible.

Our goal is to help you reach your goal—and then some. We want you to get the best score possible on any test of reading comprehension; and we also want you to be able to read with enough speed and understanding so that your studying time is cut in half, and your pleasure reading time is multiplied.

You *can* upgrade your reading ability—but you must have a plan—a procedure—a method. First, let us understand that there are two aspects of success in reading interpretation:

1. READING SPEED

 and

2. UNDERSTANDING WHAT YOU READ.

But these two aspects are not separate. As a matter of fact, they are totally dependent on each other. You can improve your speed by improving your comprehension—and then your comprehension will improve further because you have improved your speed. What you are improving, therefore, is your *speed of comprehension.* Your eyes and your mind must work together. As your mind begins to look for ideas rather than words, your eyes will begin to obey your mind. Your eyes will start to skim over words, looking for the ideas your mind is telling them to search for. Good reading is good thinking—and a good thinker will be a good reader. Speed and comprehension work together.

For convenience, however, let us divide our discussion into two parts—increasing reading speed and improving reading comprehension.

Increasing Reading Speed

A great many people read very slowly and with little comprehension, yet are completely unaware of just how badly they do read. Some people pronounce the words to themselves as they read, saying each word almost as distinctly as though reading aloud; or they think each one separately.

The reason for this is that many people have not gone quite far enough in their "learning to read" process. When you were first taught to read, you learned the sounds of each letter. Then you learned that if you put the letters together, they would make words. But that is where many people stop. Reading, to them, is reading words. But try reading a sentence out loud, saying each word as though it were a separate unit. How does it sound? Pretty meaningless! A more mature reader will put words together to make phrases. And the most mature reader will put phrases together to make ideas. A writer uses words to state ideas—and that is what a good reader looks for—those ideas. This will affect the way his eyes work. Let's see how.

HOW YOUR EYES WORK IN READING

As you learn to read phrases and thoughts, you will find that your eyes are increasing their *span.* This means that your eyes are seeing several words at a time as you are reading, not just one.

Your eyes work as a camera does. When you want to take a picture, you hold the camera still and snap the shutter. If you move the camera, the picture will blur. When you read, your eyes take pictures of words—and, like a camera, when they are "photographing," they are standing still. Each time your eyes "picture" words in a line of print, they stop—and each stop is called a *fixation*. Watch someone read, and you will see how his eyes make very quick stops across the line. You know he has finished a line when you see his eyes sweep back to the beginning of the next line.

EYE SPAN AND FIXATION

The more words your eyes take in with one fixation, the larger the eye span. And the larger the eye span, the fewer stops your eyes will have to make across the line. Thus, you will be reading faster.

For example, let's divide a sentence the way a slow, word-by-word reader would:

> You/ will/ find/ that/ you/ can/ read/ faster/ if/ you/ per/ mit/ your/ eyes/ to/ see/ large/ thought/ units/.

The reader's eyes have made at least nine stops on each line.

This is the way a fast reader would divide the same sentence:

> You will find that/ you can read faster/ if/ you permit/ your eyes to see/ large thought/ units.

This reader's eyes have stopped only three times on each line, so of course he will be able to read much faster. Also, reading thought units will enable him to grasp the meaning more effectively. Now here are some exercises to help you increase your eye span.

EXERCISES FOR INCREASING EYE-SPAN

1.

0.................0.................0.................0
0.................0.................0.................0
0.................0.................0.................0
0.................0.................0.................0
0.................0.................0.................0
0.................0.................0.................0

In the above "paragraph," the dots stand for letters and each 0 is one eye fixation. "Read" a line, forcing your eyes to shift from 0 to 0. When you finish the first line, let your eyes swing back to the next line. Try to get an even rhythm. Now you can feel what your eyes should be doing as they read a line in four fixations. Is this different from the way they usually feel when you read? Keep practicing this "paragraph" until it feels comfortable, and then try to read a line of print in the same way. You can make up your own "paragraph" with only three fixations and practice.

2. Here is a list of three-word phrases, with a line drawn down the center. Focus your eyes on the line, and look at the three words at once. Remember, only *one* fixation. Do not read each word separately. If you have trouble at first, read the phrases through once in your usual way, and then practice the one fixation.

at the store
day and night
box of candy
come with me
in the house
bring my paper
time to finish
make every effort
all the questions
read very fast

3. Choose a newspaper column on a subject that interests you, and read it through. Then draw two vertical lines equally distant from each other down the center. Reread the column fixating on first one vertical line, then the next—two fixations per line. When you get very good at this, try drawing just one line down the center and fixating once. You can practice this daily.

VOCALIZING CAN SLOW YOU DOWN

Some readers move their lips or whisper while they read "silently." This habit is called vocalizing. It is caused by the fact that your earliest reading was done aloud, and the habit of hearing each word as you read it, persists.

It would be physically impossible for you to speak at the rate of speed at which a good reader can read—say 350 words per minute. And if you could, no one could understand you. If you read only as fast as you can talk, you will never be a fast reader.

Obviously, then, you must stop vocalizing. Your

lips and vocal cords must not be permitted to interfere in the exchange of ideas between eyes and mind. Even if you are not obviously vocalizing, you may be subvocalizing. Your lips are not moving, your vocal cords are not involved, but you are hearing each word as you read it. This is as much a deterrent to reading speed as actual vocalizing. Most people *do* subvocalize.

HOW TO STOP VOCALIZING

1. Put your fingers on your lips. Make sure your lips do not move as you read. If they do, put a pencil or a rubber eraser between your teeth. Then read. If you start to vocalize, the pencil will drop out. If you are reading in public you might be embarrassed to appear with a pencil in your mouth. In that case, just clench your teeth hard—and keep reading.

2. Only *you* will know if you are subvocalizing—and be honest with yourself. If you are subvocalizing (and you probably are) try this exercise. Before you start to read, repeat these nonsense syllables to yourself for 30 seconds: da-rum, da-rum, da-rum, da-rum, etc. Now begin to read and continue to repeat da-rum as you read. If you are doing this, then you cannot subvocalize what you are reading. At first you will find this extremely difficult to do, but if you keep practicing, soon you will find that there is a direct connection between written word and thought—with no intervening vocalizing.

VARY YOUR READING SPEED

One should adjust his reading speed to what he is reading. Some paragraphs will be easier for you than others, possibly because you are more interested in the subject matter, or know something about it. Other paragraphs, particularly those that deal with factual or technical material, may have to be read more slowly.

Flexibility should be employed so that the reader will change his speed from paragraph to paragraph—even from sentence to sentence, just as a driver would vary his driving speed depending on where he is driving. Some passages are open highways while others are crowded city thoroughfares.

For example, read the following passage:

> It was a sunny Sunday afternoon in December. Some people were at the movies; some were out walking; and some were at home listening to the radio. Suddenly an announcement was broadcast—and the United States was plunged into war.
>
> On December 7, 1941, the Japanese Air Force attacked Pearl Harbor, destroying battleships, aircraft carriers, planes, and a strategic military base, leaving the United States without the military arsenals needed for anti-aircraft activity and civilian protecton.

Which of these paragraphs is the "highway"? Which is the crowded city thoroughfare? Where can you breeze through? Where will you need to slow down to absorb every detail. You're right! The first paragraph is a simple introduction. A glance should suffice. The second paragraph is fact-packed, so you will need to slow down.

OTHER PHYSICAL FACTORS

Don't neglect the obvious reading aids. Good eyesight is essential. When was your last eye check-up? If glasses were prescribed, are you using them? Make sure that you are physically comfortable, sitting erect wlth head slightly inclined. You should have good direct and indirect light, with the direct light coming from behind you and slightly above your shoulder. Hold your reading matter at your own best reading distance so that you don't have to stoop or squint.

FORCE YOURSELF TO FASTER READING

Now that you know the elements that make for fast reading, you must continue to force yourself to read as quickly as you can. Use a stop-watch to time yourself. You can figure your rate of speed by dividing the number of words on a page into the number of seconds it took you to read it, and then multiplying by 60. This will give you your rate in words-per-minute. Since no one rate of speed is possible for all reading material, your rate will vary. But an average reading speed of 350 words-per-minute should be possible for uncomplicated, interesting, straightforward material. If you are already reading that fast, then try for 500 words-per-minute. You should be able to answer correctly at least 80 per cent of the questions following a reading passage.

Practice reading quickly. Move your eyes rapidly across the line of type, skimming it. Don't permit your eyes to stop for individual words. Proceed quickly through the paragraph without backtracking. If you think you don't understand what you are reading, then reread two or three times—but always read quickly. You will be amazed to discover how much you actually do understand.

Improving Reading Comprehension

Many readers are afraid of not understanding what they read quickly. But the old idea that slow readers make up for their slowness by better comprehension of what they read has been proven untrue. Your ability to comprehend what you read will keep pace with your increase in speed. You will absorb as many ideas per page as before, and get many more ideas per unit of reading time.

It has been demonstrated that those who read quickly also read best. This is probably due to the fact that heavier concentration is required for rapid reading; and concentration is what enables a reader to grasp important ideas contained in the reading material.

GETTING THE MAIN IDEA

A good paragraph generally has one central thought—and that thought is usually stated in one sentence. That sentence, the *topic sentence,* is often the first sentence of the paragraph, but it is sometimes buried in the middle, or it can be at the end. Your main task is to locate that sentence and absorb the thought it contains while reading the paragraph. The correct interpretation of the paragraph is based on that thought *as it is stated,* and not on your personal opinion, prejudice, or preference about that thought.

Here are several examples of paragraphs. Read them quickly and see if you can pick out the topic sentence. It is the key sentence. The rest of the paragraph either supports or illustrates it. The answers follow the paragraphs.

1. Pigeon fanciers are firmly convinced that modern inventions can never replace the carrier pigeon. "A pigeon gets through when everything else fails," they say. In World War II, one pigeon flew twenty miles in twenty minutes to cancel the bombing of a town. Radios may get out of order and telephone lines may get fouled up, but the pigeon is always ready to take off with a message.
2. When a piece of paper burns, it is completely changed. The ash that is left behind does not look like the original piece of paper. When dull-red rust appears on a piece of tinware, it is quite different from the gleaming tin. The tarnish that forms on silverware is a new substance unlike the silver itself. Animal tissue is unlike the vegetable substance from which it is made. A change in which the original substance is turned into a different substance is called a chemical change.
3. A child who stays up too late is often too tired to be successful in school. A child who is allowed to eat anything he wishes may have bad teeth and even suffer from malnutrition. Children who are rude and disorderly often suffer pangs of guilt. Children who are disciplined are happy children. They blossom in an atmosphere where they know exactly what is expected of them. This provides them with a sense of order, a feeling of security.

Answers: In paragraph 1, the first sentence is the topic sentence. In paragraph 2, the last sentence is the topic sentence. In paragraph 3, it is the fourth sentence—"Children who are disciplined. . . . ".

If a selection consists of two or more paragraphs, the correct interpretation is based on the central idea of the entire passage. The ability to grasp the central idea of a passage can be acquired by practice—practice that will also increase the speed with which you read.

Reading for a Purpose—The Survey Method

Many readers don't know what they are looking for when they read. They plunge into a page full of words, and often that is what they end up with —just words. It's like walking into a supermarket without having made a list of what you want to buy. You wander aimlessly up and down the aisles and end up with a basketful of cookies and fruit and pickles—and nothing for a main dish.

It is extremely important to have a purpose in mind *before* you start to read—to make a "list" before you start shopping. Good readers use the *survey* method. By "survey" we mean a quick over-view of what you are going to read before you actually start reading. It is like looking at a road map before you start on a trip. If you know in what direction you are going, you are apt to get there sooner, and more efficiently. This is what you do.

1. Read the title. Think about what the selection will probably be about. What kind of information can you expect to obtain? Gear your mind to look for the central thought.

2. Think about the kind of vocabulary you will meet. Will it be technical? Are you familiar with the subject, or will you have to prepare yourself to meet many new words? After a quick glance, you may decide to skip this selection and go back to it later. (Remember, on a timed reading test you want to give yourself a chance to sample *all* the selections. The one at the end may be easier for you than the one in the middle, but you won't know if you never get to the end.) The difficulty of the vocabulary may be the deciding factor.

3. If there are subheadings, read them. They can provide a skeleton outline of the selection.

4. Read the first sentence of each paragraph. It usually contains the most important ideas in the selection. The topic sentence is more often found in the beginning of a paragraph than in any other position.

5. READ THE QUESTIONS BASED ON THE SELECTION. The questions are there to test whether or not you understand the most important ideas in the selection. If you read them first, they will steer you through your reading in the most effective way possible. Now you really know what to look for! In any kind of reading, whether on test or in texts, always look at the questions first (unless you are directed not to do so).

The survey method can be applied to all kinds of reading, particularly textbook reading. In addition to the above, you should include the following in your textbook survey:

(a) Read the preface quickly. It states the author's purpose in writing the book.

(b) Look at the publication date (on the copyright page). This can tell you if the information is up-to-date.

(c) Look through the table of contents. See what the author has included, and the order in which it appears. Some tables of contents can serve as an outline for the book.

(d) If there are chapter or part summaries, read through them quickly. They'll give you a forecast of what's to come.

(e) Look at illustrations, maps, graphs, etc. These are meant to help you visualize essential information. Remember, one picture can be worth a thousand words.

Cues and Clues For Readers

Examination points may be unnecessarily lost by ignoring the author's hints as to what *he* thinks is most important. Be on the lookout for such phrases as "Note that . . ." "Of importance is . . ." "Don't overlook . . ." These give clues to what the writer is stressing. Beware of negatives and all-inclusive statements. They are often put in to trip you up. Words like *always, never, all, only, every, absolutely, completely, none, entirely, no,* can sometimes turn a reasonable statement into an untrue statement. For example look at the following sentence:

When you get caught in the rain, you catch cold.

True? Of course. Now look at this sentence:

When you get caught in the rain, you *always* catch cold.

Different, isn't it? Not *always* true.

PUNCTUATION

Other hints which you should also watch for are those given by punctuation. Here are a few points to keep in mind:

1. QUOTATION MARKS—When a statement is quoted, it may not necessarily represent the author's opinion, or the main thought of the passage. Be sure you make this distinction if it is called for.

2. EXCLAMATION POINT—This mark is often used to indicate an *emphatic* or *ironical* comment. It's the author's way of saying, "This is important!"

3. COMMAS—Watch those commas. They can change the meaning of a sentence. For example:

As I left the room, in order to go to school John called me.

As I left the room in order to go to school, John called me.

In each sentence, a different person is going to school.

4. PARENTHESES—These are often used to set off a part of the sentence that is not absolutely necessary to the sentence. But don't ignore them in reading comprehension tests. Sometimes they give vital information. For example:

Shakespeare (whose life spanned the sixteenth and seventeenth centuries) was a great dramatist.

5. COLON—Often used to emphasize a sequence in thought between two independent sentences. For example:

Science plays an important role in our civilization: thus we should all study physics, chemistry, and biology.

6. ELLIPSES—Three dots often found in quoted material which indicates that there has been an omission of material from the original quotation. Often the material omitted is not important, but a good, critical reader should be aware of the omission.

A Systematic Plan

You can't sit down the night before a test in reading comprehension and "cram" for it. The only way you can build up your reading skill is to practice systematically. The gains you make will show up not only in an increased score on a reading comprehension test, but also in your reading for study and pleasure.

Trying to change reading habits that you have had for a long time can be difficult and discouraging. Do not attempt to apply *all* of the suggestions we have given to *all* your reading *all* at once. Try to follow a program like the one below.

1. Set aside 15 minutes a day to practice new reading techniques.

2. Start off with a short, easy-to-read article for a magazine or newspaper—and time yourself. At the end of your practice session, time yourself on another short article, and keep a record of both times.

3. Select a news story. Read it first, and then practice an eye-span exercise. Work towards reducing your eye fixations to no more than two for a line, the width of a newspaper column.

4. Read an editorial, book review, or movie or drama review in a literate magazine or newspaper. This type of article always expresses the author's (or the paper's) point of view and is therefore good practice for searching out the main idea. After you read, see whether you can write a good title for the article and jot down in one sentence the author's main idea. Also, you can try making up a question based on the article with five alternate answers (the kind you find on reading comprehension tests). This is excellent practice for determining main ideas, and you can use the questions to test your friends.

5. Find one new word and write the sentence in which it appears. Guess at its meaning from the context. Then look up the definition in a dictionary and try to make up a sentence of your own, using the word. Then try to use the word in your conversation at least twice the following day.

If you follow this program daily, you will soon find that you can extend to more and more reading the skills you are building, and your reading comprehension test score will show the great gains you have made.

TEN SUCCESS STEPS

Here are proven techniques for getting the right answer to *any* Reading Interpretation question.

Survey Selection

1. Read the selection through quickly to get the general sense.
2. Reread the selection, concentrating on the central idea.
3. Can you now pick out the *topic sentence* in each paragraph?
4. If the selection consists of more than one paragraph, determine the *central idea* of the entire selection.

Survey Stems
Concentrate on Each Question

5. Examine the five choices carefully, yet rapidly. Eliminate immediately those choices which are *far-fetched, ridiculous, irrelevant, false,* or *impossible.*
6. Eliminate those choices which may be true, but which have nothing to do with the sense of the selection.
7. Check those few choices which now remain as possibilities.

Reread Selectively
Shuttle Back to Selection

8. Refer back to the original selection and determine which one of these remaining possibilities is best in view of

a) specific information in the selection
or
b) implied information in the selection

Reread only the part of the selection that applies to the question, and make your decision as to the correct choice based on these considerations:

(a) A choice must be based on fact actually given or definitely understood (and not on your personal opinion or prejudice.) Some questions require making a judgment—and this judgment also must be based on the facts as given.

(b) In questions involving the central thought of the passage (for example: "The best title for this selection . . .") the choice must accurately reflect the entire thought—not too narrow, and not too general.

9. Be sure to consider only the facts *given* or *definitely understood* some place in the selection.

10. Be especially careful of trick expressions or "catch-words" which sometimes destroy the validity of a seemingly acceptable answer. These include the expressions: "under all circumstances," "at all times," "never," "always," "under no conditions," "absolutely," "completely," and "entirely."

READING COMPREHENSION FOR FIREFIGHTER

Questions numbered 1 through 4 relate to the paragraph below and are to be answered in accordance with the paragraph.

"The unadjusted loss per $1000 valuation has only a very limited usefulness in evaluating the efficiency of a fire department, for it depends upon the assumption that other factors will remain constant from time to time and city to city. It might be expected that high fire department operation expenditures would tend to be associated with a low fire loss. A statistical study of the loss and cost data in more than 100 cities failed to reveal any such correlation. The lack of relationship, although to some extent due to failure to make the most efficacious expenditure of fire protection funds, must be attributed in part at least to the obscuring effect of variations in the natural, physical, and moral factors which affect fire risk."

1. One reason for the failure to obtain the expected relationship between fire department expenditures and fire loss data which is stated in the above paragraph is
   (A) the changing dollar valuation of property
   (B) the unsettling effects of rapid technological innovations
   (C) the inefficiency of some fire department activities
   (D) the statistical errors made by the investigators.
2. We may conclude that the "unadjusted loss per $1000" figure is useful in comparing the fire departments of two cities
   (A) only if the cities are of comparable size
   (B) only if adjustments are made for other factors which affect fire loss
   (C) under no circumstances
   (D) only if properly controlled experimental conditions can be obtained.
3. The one of the following factors which affect fire risk that is most adequately reflected in the "unadjusted loss per $1000 valuation" index is
   (A) fire department operation expenditures
   (B) physical characteristics of the city
   (C) type of structures most prevalent in the city
   (D) total worth of property in the city.
4. According to the above paragraph, cities which spend larger sums on their fire departments
   (A) tend to have lower fire losses than cities which spend smaller sums on their fire departments
   (B) do not tend to have lower fire losses than cities which spend smaller sums on their fire departments
   (C) tend to have higher fire losses than cities which spend smaller sums on their fire departments
   (D) do not tend to have the same total property valuation as cities which spend smaller sums on their fire departments.

Questions numbered 5 through 7 relate to the paragraph below and are to be answered in accordance with the paragraph.

"Shafts extending into the top story, except those stair shafts where the stairs do not continue to the roof, shall be carried through and at least two feet above the roof. Every shaft extending above the roof, except open shafts and elevator shafts, shall be enclosed at the top with a roof of materials having a fire resistive rating of one hour and a metal skylight covering at least three-quarters of the area of the shaft in the top story, except that skylights over stair shafts shall have an area not less than one-tenth the area of the shaft in the top story, but shall be not less than fifteen square feet in area. Any shaft terminating below the top story of a structure and those stair shafts not required to extend through the roof shall have the top enclosed with materials having the same fire resistive rating as required for the shaft enclosure."

5. The above paragraph states that the elevator shafts which extend into the top story
   (A) are not required to have a skylight but are required to extend at least two feet above the roof
   (B) are neither required to have a skylight nor to extend above the roof
   (C) are required to have a skylight covering at least three-quarters of the area of the shaft in the top story and to extend at least two feet above the roof
   (D) are required to have a skylight covering at least three-quarters of the area of the shaft in the top story but are not required to extend above the roof.
6. The one of the following skylights which meets the requirements of the above paragraph is a skylight measuring
   (A) 4′ x 4′ over a stair shaft which, on the top story, measures 20′ x 9′

(B) $4\frac{1}{2}'$ x $3\frac{1}{2}'$ over a pipe shaft which, on the top story, measures 5′ x 4′
(C) $2\frac{1}{2}'$ x $1\frac{1}{2}'$ over a dumb-waiter shaft which, on the top story, measures $2\frac{1}{2}'$ x $2\frac{1}{2}'$
(D) 4′ x 3′ over a stair shaft which, on the top story, measures 15′ x 6′.

7. Suppose that in a Class I building a shaft which does not go to the roof is required to have a three hour fire resistive rating. In regard to the material enclosing the top of this shaft, the above paragraph
(A) states that a one hour fire resistive rating is required
(B) states that a three hour fire resistive rating is required
(C) implies that no fire resistive rating is required
(D) neither states nor implies anything about the fire resistive rating.

The following four questions, numbered 8 to 11 inclusive, relate to the information given in the following paragraph and are to be in accordance with this information.

"Old buildings are liable to possess a special degree of fire risk merely because they are old. Outmoded electrical wiring systems and installation of new heating appliances for which the building was not designed may contribute to the increased hazard. Old buildings have often been altered many times; parts of the structure may antedate building codes; dangerous defects may have been covered up. On the average, old buildings contain more lumber than comparable new buildings which, in itself, makes old buildings more susceptible to fire. It is not true, though, that sound lumber in old buildings is drier than new lumber. Moisture content of lumber varies with that of the atmosphere to which it is exposed."

8. According to the above paragraph, old buildings present a special fire hazard chiefly because of the
(A) poor planning of the buildings when first designed
(B) haphazard alteration of the buildings
(C) failure to replace worn out equipment
(D) inadequate enforcement of the building codes.

9. We may conclude from the above paragraph that lumber
(A) should not be used in buildings unless absolutely necessary
(B) should not be used near electrical equipment
(C) is more inflammable than newer types of building materials
(D) tends to lose its moisture at a constant rate.

10. According to the above paragraph, the amount of moisture in the wooden parts of a building depends upon the
(A) age of the building
(B) moisture in the surrounding air
(C) type of heating equipment used in the building
(D) quality of lumber used.

11. In regard to building codes, the above paragraph implies that
(A) old buildings are exempt from the provisions of building codes
(B) some buildings now in use were built before building codes were adopted
(C) building codes usually don't cover electrical wiring systems
(D) building codes generally are inadequate.

The following three questions, numbered 12 to 14 inclusive, relate to the information given in the following paragraph and are to be answered in accordance with this information.

"Fire records indicate that about 10 per cent of the total value of fire loss in the United States is due to fires caused by heating equipment, and that insufficient clearance to combustible materials is an important factor in a large percentage of these fires. The need for adequate clearance is emphasized at periods of severe cold weather, when heating equipment is run at full capacity, by a marked increase in the number of fires."

12. In a given year in the United States 2,000,000 fires caused 10,000 deaths and destroyed $500,000,000 worth of property. We may conclude from the above paragraph that heating equipment caused
(A) 200,000 fires
(B) fires resulting in the death of 1,000 persons
(C) fire resulting in the destruction of $50,000,000 worth of property
(D) 200,000 fires resulting in the death of 1,000 persons and the destruction of $50,000,000 worth of property.

13. By "clearance to combustible materials," as used in the above paragraph, is meant
(A) testing laboratory approval for heating equipment
(B) maintaining heating equipment in clean condition

(C) keeping sufficient space between heating equipment and inflammable material
(D) obtaining appropriate permits before installing the heating equipment.

14. The month-to-month changes in the number of fires caused by heating equipment is most affected by the
(A) season of the year
(B) amount of clearance to combustible materials
(C) condition of the equipment
(D) long term trend in fire statistics.

"Division commanders shall arrange and maintain a plan for the use of hose wagons to transport members in emergencies. Upon receipt of a call for members, the deputy chief of the division from whom the men are called shall have the designated hose wagon placed out of service and prepared for the transportation of members. Hose wagons shall be placed at central assembly points, and members detailed instructed to report promptly to such locations equipped for fire duty. Hose wagons designated shall remain at regular assignments when not engaged in the transportation of members."

15. Preparation of the hose wagon for this special assignment of transporting of members would most likely involve
(A) checking the gas and oil, air in tires and mechanical operation of the apparatus
(B) removal of hose lines to make room for the members being transported
(C) gathering of equipment which will be needed by the members being transported
(D) instructing the driver on the best route to be used.

16. Hose wagons used for emergency transportation of members are placed out of service because they
(A) are not available to respond to alarms in their own district
(B) are more subject to mechanical breakdown while on emergency duty
(C) are engaged in operations which are not the primary responsibility of their division
(D) are considered reserve equipment.

17. Of the following, the best example of the type of emergency referred to in the above paragraph is
(A) a fireman injured at a fire and requiring transportation
(B) a subway strike which prevents firemen from reporting for duty
(C) an unusually large number of false alarms occurring at one time
(D) a need for additional manpower at a fire.

18. A "central assembly point," as used in the above paragraph, would most likely be a place
(A) close to the place of the emergency
(B) in the geographical center of the division
(C) easily reached by the members assigned
(D) readily accessible to the intersection of major highways.

The following four questions, numbered 19 to 22 inclusive, are based on the information given in the paragraph below and are to be answered in accordance with this information.

"The principal value of inspection work is in the knowledge obtained relating to the various structural features of the building and the protective features provided. Knowledge of the location of stairways and elevators, the obstruction provided by merchandise, the danger from absorption of water by baled stock, the potential hazard of rupture of containers such as drums or cylinders, and the location of protective equipment, all are essential features to be noted and later discussed in company school and officer's college."

19. According to the above paragraph, the chief value of inspection work is to gather information which will aid in
(A) fixing responsibility for fires
(B) planning fire fighting operations
(C) training new firemen
(D) obtaining compliance with the Building Code.

20. The one of the following objects which would be the most help in accomplishing the objective of the inspection as stated in the above paragraph is a
(A) copy of the Building Code
(B) chemical analysis kit
(C) plan of the building
(D) list of the building's tenants.

21. An example of a "structural feature" contained in the above paragraph is the
(A) location of stairways and elevators
(B) obstruction provided by merchandise
(C) danger of absorption of water by baled stock
(D) hazard of rupture of containers such as drums or cylinders.

22. Of the following, the best example of what is meant by a "protective feature," as used in the above paragraph, is
(A) a fire extinguisher
(B) a burglar alarm

(C) fire insurance
(D) a medical first-aid kit.

23. "It used to be quite the common thing for firemen to use large streams of water to extinguish small fires, and in some cases where the contents of the buildings were not entirely ruined by fire, the damage by water exceeded the fire damage as such." This statement illustrates the fact that
(A) considerable water is vitally necessary to extinguish fires
(B) water should be available near the scene of a fire
(C) water without hose is of little use
(D) courage is a greater asset to firemen than modern apparatus
(E) fire losses are not always the consequence of flames.

Items 24 to 30 are based on the following paragraphs:

"Fire regulations require that every liquefied petroleum gas installation should be provided with the means for shutting off the supply to a building in case of an emergency. The installation of a shut-off valve immediately inside a building, which is sometimes done for the convenience of the user, does not comply with this regulation. An outside shut-off valve just outside the building seems to be the logical solution. However, the possibility of tampering illustrates the danger of such an arrangement. A shut-off valve so located might be placed in a locked box. However, this has no advantage over a valve provided within the locked cabinet containing the cylinder or an enclosure provided over the top of the cylinders. Keys may be carried by firemen or in an emergency the lock may be broken. Where no valve is visible, the firemen should not hesitate to break the lock to the cylinder enclosure. The means for shutting off the gas varies considerably in the numerous types of equipment in use. When the cover to the enclosure has been opened, the gas may be shut off as follows:

"Close the tank or cylinder valves to which the supply line is connected. Such valves always turn to the right. If the valve is not provided with a handwheel, an adjustable wrench can be used. If conditions are such that shutting off the supply at once is imperative and this cannot be accomplished as above, the tubing which is commonly employed as the supply line can be flattened to the extent of closure by a hammer. If the emergency is such as to require the removal of the cylinder, the supply line should be disconnected and the cylinder removed to a safe location. A tank buried in the ground is safe against fire. When conditions indicate the need of removing a cylinder or tank, and this cannot be done due to the severity of exposure, pressures within the container can be kept within control of the safety valve by means of a hose stream played on the surface of the container. The melting of the fuse plug may also be prevented in this way."

24. According to the preceding paragraphs, in an emergency a fireman should break the lock of a cylinder enclosure whenever the shut-off valve
(A) fails to operate
(B) has no handwheel
(C) has been tampered with
(D) cannot be seen
(E) is installed inside the building.

25. According to the preceding paragraphs, shut-off valves for liquefied petroleum gas installations
(A) always turn to the right
(B) always turn to the left
(C) sometimes turn to the right and sometimes turn to the left
(D) are generally pulled up
(E) are generally pushed down

26. According to the preceding paragraphs, if a cylinder should be moved but cannot because of the severity of exposure, the pressure can be kept under control by
(A) opening the shut-off valve
(B) playing a hose stream on the cylinder
(C) disconnecting the supply line into the cylinder
(D) removing the fuse plug
(E) closing the cabinet containing the cylinder.

27. The preceding paragraphs state that the supply line should be disconnected when the
(A) fuse plug melts
(B) cylinder is removed to another location
(C) supply line becomes defective
(D) cylinder is damaged
(E) shut-off valve fails to operate.

28. The preceding paragraphs state that shut-off valves for liquefied petroleum gas installations are sometimes placed inside buildings
(A) so that firemen will be able to find the valves more easily
(B) because it is more convenient for the occupants
(C) in order to hide the valves from public view
(D) as this makes it easier to keep the valves in good working condition
(E) to insure that the valves cannot be tampered with.

29. It is suggested in the preceding paragraphs that during an emergency the supply line tubing should be flattened to the extent of closure when the
    (A) supply line becomes defective
    (B) shut-off valve cannot be opened
    (C) shut-off valve cannot be closed
    (D) supply line is very near a fire
    (E) shut-off valve is located inside the building.

30. According to the preceding paragraphs, fire regulations require that liquefied petroleum gas installations should
    (A) be made in safe places
    (B) be tamper-proof
    (C) ha e shut-off valves
    (D) not exceed a certain size
    (E) contain cylinders.

Items 31 to 40 are based on the following paragraphs:

"Air conditiong systems are complex and are made up of several processes. The circulation of the air is produced by fans and ducts; the heating is produced by steam, hot water coils, coal, gas or oil fired furnaces; the cooling is done by ice or mechanical refrigeration and the cleaning is done by air washers or filters.

"Air conditioning systems in large buildings generally should be divided into several parts with wholly separate ducts for each part or floor. The ducts are then extended through fire partitions. As a safeguard, whenever ducts pass through fire partitions, automatic fire dampers should be installed in the ducts. Furthermore, the ducts should be lined on the inside with fire-resistant materials. In addition, a manually operated fan shut-off should be installed at a location which will be readily accessible under fire conditions.

"Most air conditioning systems recirculate a considerable portion of the air and when this is done an additional safeguard has to be taken to have the fan arrange to shut down automatically in case of fire. A thermostatic device in the return air duct will operate the shut-off device whenever the temperature of the air coming to the fan becomes excessive. The air filters are frequently coated with oil to help catch dust. Such oil should be of a type that does not ignite readily. Whenever a flammable or toxic refrigerant is employed for air cooling, coils containing such refrigerant should not be inserted in any air passage."

31. According to the preceding paragraphs, fan shut-offs in the air conditioning system should be installed
    (A) near the air ducts
    (B) next to the fire partitions
    (C) near the fire dampers
    (D) where fires may start
    (E) where they can be reached quickly.

32. On the basis of the preceding paragraphs, whenever a fire breaks out in a building containing an air conditioning system which recirculates a portion of the air
    (A) the fan will shut down automatically
    (B) the air ducts will be opened
    (C) the thermostat will cease to operate
    (D) the fire partitions will open
    (E) it will be extinguished.

33. The preceding paragraphs state that on every floor of a large building where air conditioning systems are used there should be
    (A) an automatic fire damper
    (B) a thermostatic device
    (C) an air filter
    (D) a manually operated fan shut-off
    (E) a separate duct.

34. From the preceding paragraphs, the conclusion can be drawn that in an air conditioning system flammable refrigerants
    (A) may be used if certain precautions are observed
    (B) should be used sparingly and only in air passages
    (C) should not be used under any circumstances
    (D) may be more effective than other refrigerants
    (E) may be less effective than other refrigerants.

35. According to the preceding paragraphs, the spreading of dust by means of the fans in the air conditioning system is reduced by
    (A) shutting down the fan automatically
    (B) lining the inside of the air ducts
    (C) cleaning the circulated air with filters
    (D) coating the air filters with oil
    (E) cooling the circulated air with refrigerants.

36. According to the preceding paragraphs, the purpose of a thermostatic device is to
    (A) regulate the temperature of the air conditioning system
    (B) shut off the fan when the temperature of the air rises
    (C) operate the fan when the temperature of the air falls
    (D) assist the recirculation of the air
    (E) operate the air duct automatically in case of fire.

37. According to the preceding paragraphs, hot water coils in the air conditioning system are limited to
(A) cooling
(B) heating
(C) heating and cooling
(D) heating and cleaning
(E) cooling and cleaning.

38. The parts of an air conditioning system which the preceding paragraphs state should be made of fire-resistant materials are the
(A) hot water coils
(B) automatic fire dampers
(C) air duct linings
(D) thermostatic devices
(E) air filters.

39. According to the preceding paragraphs, automatic fire dampers should be installed
(A) on oil fired furnaces
(B) on every floor of a large building
(C) in ducts passing through fire partitions
(D) next to the hot water coils
(E) near the thermostatic device.

40. On the basis of the preceding paragraphs, the most accurate statement is that coils containing toxic refrigerants should be
(A) used only when necessary
(B) lined with fire-resistant materials
(C) coated with non-flammable oil
(D) kept out of any air passage
(E) kept from overheating.

Key Answers: Reading Comprehension

1. C	9. C	17. D	25. A	33. E
2. B	10. B	18. C	26. B	34. A
3. D	11. B	19. B	27. B	35 D
4. B	12. C	20. C	28. B	36. B
5. A	13. C	21. A	29. C	37. B
6. B	14. A	22. A	30. C	38. C
7. B	15. B	23. E	31. E	39. C
8. B	16. A	24. D	32. A	40. D

PREVIOUS TEST QUESTIONS FOR PRACTICE

1. A fireman who is *nimble* is best described as
(A) able to perform tasks of great importance
(B) quick and light in motion
(C) fatigued by heavy work
(D) capable of learning efficiently
(E) injured seriously.

2. "A fireman's work may sometimes be tedious." The word *tedious* means most nearly
(A) boring (B) dangerous
(C) unproductive (D) difficult
(E) solitary.

3. "The fireman was reproved for his action." The word *reproved* means most nearly
(A) praised highly (B) censured mildly
(C) rewarded (D) cited publicly
(E) exonerated.

4. "When at a fire, the fireman must avoid any action that is superfluous." The word *superfluous* means most nearly
(A) irresponsible (B) uncertain
(C) uncoordinated (D) risky
(E) unnecessary.

5. Fumes which are *toxic* are best described as
(A) dense (B) concentrated
(C) poisonous (D) offensive
(E) odorous.

6. "Officers will submit a report of all obsolete equipment." The word *obsolete* means most nearly
(A) failing to operate (B) difficult to replace
(C) out of date (D) unusually effective
(E) highly expensive.

7. "Firemen performing inspection duty should note any conditions that are hazardous." The word *hazardous* means most nearly
(A) negligent (B) congested
(C) illegal (D) dangerous
(E) unusual.

8. "The principle of operation of a fire pump is complex." The word *complex* means most nearly
(A) important (B) uniform
(C) universal (D) intricate
(E) fundamental.

9. A fireman who is *intrepid* is, most nearly,
(A) prudent (B) competent
(C) fearless (D) responsible
(E) intelligent.

10. Instructions which are *tentative* in nature are best described as
(A) precisely stated
(B) subject to change
(C) detailed in form
(D) very clear
(E) especially meaningful.

11. "Possibility of the collapse of the building was remote." The word *remote* means most nearly
(A) slight (B) imminent
(C) real (D) serious
(E) obvious.

12. To say that evidence of arson is *inconclusive* means most nearly that the evidence
(A) is entirely reasonable
(B) has been gathered irregularly
(C) is irrelevant and immaterial
(D) fails to establish definite proof
(E) has not been subjected to tests.

13. A fireman who is *meticulous* in his work is, most nearly,
(A) alert and cooperative
(B) unusually susceptible to errors
(C) courageous in emergencies
(D) a rapid but careless worker
(E) excessively careful in small details

14. "The report on the accident involving the fire apparatus was written tersely." The word *tersely* means most nearly
(A) concisely (B) competently
(C) immediately afterwards (D) poorly
(E) in accordance with instructions.

15. "Reports should include all pertinent material." The word *pertinent* means most nearly
(A) directly observed by the writer
(B) required by rules, procedure, or law
(C) having to do with the subject being considered
(D) tending to fix responsibility
(E) requested by one's superior.

16. "The members of the company extricated the fireman." As used in this sentence, the word *extricated* means most nearly
(A) revived (B) complimented
(C) freed (D) supported
(E) criticized.

17. "The fire of rags and paper was quenched in a short time." As used in this sentence, the word *quenched* means most nearly
(A) burned out (B) kindled
(C) extinguished (D) started
(E) diverted.

18. "The structure had a large number of windows exposed to radiant heat across a narrow alley." As used in this sentence, the word *radiant* means most nearly
(A) emitted (B) absorbed
(C) dense (D) dangerous
(E) visible.

19. "Oil fires can propagate in many ways." As used in this sentence, the word *propagate* means most nearly to
(A) destroy (B) spread
(C) burn (D) be extinguished
(E) be controlled

20. "A study was made of the efficiency of personnel and equipment under simulated air raid conditions." As used in this sentence, the word *simulated* means most nearly
(A) activated (B) difficult
(C) reduced (D) differentiated
(E) pretended.

21. "A man who builds with stone, brick, or similar materials" is called a (an)
(A) artisan (B) journeyman
(C) laborer (D) mason.

22. "A situation involving choices between equally unsatisfactory alternatives" is called a
(A) crisis (B) deadlock
(C) dilemma (D) farce.

23. "To make rows of small holes through a substance" defines the word
(A) penetrate (B) perforate
(C) permeate (D) pulsate.

24. "Gradual decrease in the width of an elongated object" is called a
(A) bevel (B) slope
(C) spiral (D) taper.

25. "Sphere of authority" is called
(A) constituency (B) dictatorial
(C) jurisdiction (D) vassal.

26. "To swing backward and forward" defines the word
(A) alternate (B) gesticulate
(C) oscillate (D) procrastinate.

27. "The pump faults were rectified." As used in this sentence, the word *rectified* means most nearly
(A) worsened (B) overlooked
(C) corrected (D) discovered
(E) recorded.

28. "Obstacles which might interfere with quick egress during a fire should be removed." As used in this sentence, the word *egress* means most nearly
(A) movement (B) progress
(C) assistance (D) communication
(E) departure.

29. "Carbon tetrachloride is a volatile liquid." As used in this sentence, the word *volatile* means most nearly
(A) having a high boiling point
(B) having a low freezing point
(C) evaporating rapidly
(D) poisonous
(E) odorous.

30. "The order was rescinded." As used in this sentence, the word *rescinded* means most nearly
(A) repeated (B) cancelled
(C) verified (D) relayed
(E) executed.

31. "Some of the training exercises are very intricate." As used in this sentence, the word *intricate* means mostly nearly
(A) intense (B) educational
(C) effective (D) useless
(E) complicated.

32. "Few states attempt to compile vital statistics on fires and fire losses." As used in this sentence, the word *compile* means most nearly
(A) collect (B) distribute
(C) conceal (D) interpret
(E) simplify.

33. "The heat waves and explosions compelled the fire fighters to operate on the periphery of the fire." As used in this sentence, the word *periphery* means most nearly the
(A) outer boundary (B) center
(C) coolest part (D) hottest part
(E) most dangerous part.

34. "The equipment was dismantled." As used in this sentence, the word *dismantled* means most nearly
(A) partly salvaged (B) cleaned thoroughly
(C) taken apart (D) repaired
(E) replaced.

35. "Under stringent regulations, persons cannot shirk their responsibilities." As used in this sentence, the word *stringent* means most nearly
(A) explicit (B) severe
(C) adjustable (D) legal
(E) equitable.

36. "A lack of buoyancy in the smoke made fighting the cellar fire more difficult." As used in

this sentence, the word *buoyancy* means most nearly having the ability to
(A) reflect light (B) rise
(C) absorb moisture (D) cool
(E) reflect heat.

37. "These were the alleged causes for the fire. As used in this sentence, the word *alleged* means most nearly
(A) known (B) proved
(C) secret (D) declared
(E) real.

38. "The result of the action was that the smoke nuisance was eradicated." As used in this sentence, the word *eradicated* means most nearly
(A) changed (B) discovered
(C) increased (D) abolished
(E) blamed.

39. "The problem was aggravated by these new actions." As used in this sentence, the word *aggravated* means most nearly
(A) solved (B) publicized
(C) revealed (D) increased
(E) lightened.

40. The difference between non-inflammable and incombustible substances is
(A) that the latter explode more readily
(B) that the latter have specific gravities near zero
(C) that the latter combine more quickly with gases
(D) that the latter respond more readily to chemical action
(E) none of the foregoing.

41. A volatile liquid is one which
(A) vaporizes readily
(B) resists oxidation
(C) heats very slowly
(D) is not a compound
(E) extinguishes fires.

42. "The wire connecting the two terminals must be kept taut." As used in this sentence, the word *taut* means most nearly
(A) without defects
(B) without slack
(C) without electrical charge
(D) without pressure.

43. "The facades of the building were dirty and grimy." As used in this sentence, the word *facades* means most nearly
(A) cellars (B) fronts
(C) residents (D) surroundings.

44. "Several firemen were injured by the detonation." As used in this sentence, the word *detonation* means most nearly
(A) accident (B) collapse
(C) collision (D) explosion.

45. To say that the decreasing fire loss in one city is an index of the efficiency of the Fire Department is to use *index* as signifying
(A) indentation (B) importance
(C) convention (D) indicator
(E) fire.

46. The fireman whose position is *precarious* is in a position which is
(A) enviable (B) high
(C) low (D) dangerous
(E) unchanged.

47. A fireman who is *lauded* for his work receives
(A) monetary reward (B) medication
(C) condemnation (D) directions
(E) praise.

48. A fireman who encounters the word *amorphous* in his reading knows that the word means
(A) without malice
(B) without determinate shape
(C) fire retarding
(D) with explicit intent
(E) chemically active.

49. Fire fighting equipment which is *superlatively* efficient is efficient to
(A) a slight degree
(B) a moderate degree
(C) a marked degree
(D) a trivial extent
(E) an intemperate extent.

50. To say that intelligence is a *requisite* for the fire fighter means that intelligence is, for the fireman,
(A) incandescent (B) indispensable
(C) incisive (D) illicit
(E) incidental.

51. "To train firemen in the handling of fire fighting implements increases the efficiency of the firemen." *Implements* means most nearly
(A) tools (B) vapors
(C) companies (D) implications
(E) impositions.

52. "Chemical extinguishers have become universally popular." *Universally* means most nearly
(A) without cause
(B) for preventive purposes
(C) chemically effective

(D) without exception
(E) dependent upon use.

53. The fireman whose reports have the virtue of *brevity* writes
(A) concisely (B) clearly
(C) forcefully (D) unintelligibly
(E) coherently.

54. The fireman who is *discreet* in his behavior is
(A) discourteous (B) disinterested
(C) prudent (D) uninterested
(E) determined.

55. The fireman who mounts a ladder in *agile* fashion is climbing
(A) nimbly (B) laboriously
(C) carefully (D) carelessly
(E) recklessly.

56. Those who achieve *eminence* in the field of fire fighting are
(A) famous (B) emblematic
(C) inert (D) generous
(E) kindly.

57. The fireman whose relationships with his fellows are *amicable* behaves in a manner which is characterized by
(A) antagonism (B) a friendly spirit
(C) self reliance (D) uncertainty
(E) caution.

58. The fireman who encounters the word *torpid* in his reading knows that the word means
(A) anomalous (B) egregious
(C) altisonant (D) homogeneous
(E) dormant.

Each of the following numbered words is followed by four suggested words. In each item, one of the suggested words or phrases has a meaning which is almost the same or almost the opposite to the meaning of the numbered word. Blacken the space between the pair of lines lettered like that of the word which is most nearly the same or most nearly the opposite in meaning to that of the numbered word.

59. corollary
(A) conundrum (B) program
(C) consequence (D) plan

60. eleemosynary
(A) philanthropic (B) divergent
(C) droll (D) handicap

61. expropriate
(A) protrude (B) erode
(C) expunge (D) deprive

62. fruition
(A) futility (B) realization
(C) resentment (D) economy

63. inception
(A) origin (B) trickery
(C) riot (D) confinement

64. reticent
(A) opulent (B) loquacious
(C) reliable (D) persuasive

65. stricture
(A) bite (B) creak
(C) surprise (D) criticism

66. subjugate
(A) conquer (B) dismiss
(C) allow (D) append

67. subversive
(A) support (B) playful
(C) corrupt (D) averse

68. tangible
(A) inflammable (B) inferior
(C) responsible (D) palpable

69. lenity
(A) cruelty (B) insanity
(C) ease (D) nimbleness

70. furtive
(A) ardent (B) stealthy
(C) gaping (D) glowing

DIRECTIONS: From each of the following groups of four words, select the word whose meaning is most nearly the OPPOSITE of the leading word.

71. acrimony
(A) wisdom (B) amiability
(C) sharpness (D) objectivity

72. adulatory
(A) imitative (B) carping
(C) helpful (D) impious

73. amenity
(A) dexterity (B) affability
(C) harshness (D) viciousness

74. apostate
(A) zealot (B) proponent
(C) designate (D) convert

75. desuetude
(A) optimism (B) improvement
(C) unity (D) aptitude

76. diaeresis

(A) diphthong (B) poetic foot
(C) synecdoche (D) syncope

77. disparage
(A) applaud (B) agree
(C) disperse (D) simulate

78. esoteric
(A) manifest B) uncertain
(C) secret (D) variable

79. forward
(A) fastidious (B) lazy
(C) complaint (D) stupid

80. imprecate
(A) bless (B) boast
(C) attack (D) encourage

DIRECTIONS: Below are 5 groups of paired words, numbered 81 through 85. For each group, select the one pair of words, lettered (A), (B), (C), (D), (E), which are opposite in meaning and mark the answer in the space on the right.

Example:
(A) behead—decapitate
(B) flirt—coquette
(C) valor—cowardice
(D) ejaculate—exclaim
(E) despicable—contemptible

Since valor and cowardice are opposite in meaning, the correct answer is C.

81. (A) dogmatic—positive
(B) baffle—aid
(C) eccentric—odd
(D) durable—lasting
(E) covetousness—avarice

82. (A) tedious—tiresome
(B) tenacious—tough
(C) dissemble—uncover
(D) sinister—evil
(E) reflect—contemplate

83. (A) latent—concealed
(B) abject—noble
(C) fabulous—incredible
(D) impute—count
(E) judgment—discernment

84. (A) laborious—arduous
(B) laudable—praiseworthy
(C) incongruous—inconsistent
(D) injudicious—imprudent
(E) dwindle—increase

85. (A) denounce—condemn
(B) insinuate—intimate
(C) encumbrance—help
(D) presumptuous—arrogant
(E) explicit—definite

DIRECTIONS: Following are ten pairs of words numbered 86 through 95. A pair of words may be the same or nearly the same in meaning, or may be opposite or nearly opposite in meaning, *or* may be related to each other in neither way. On the dotted line provided mark "S" if a pair of words are the same or nearly the same in meaning; "O" if they are opposite or nearly opposite in meaning; and "N" if they are related to each other in neither way.

86. pertinent—irrelevant ....
87. condone—forgive ....
88. obdurate—yielding ....
89. allocation—assertion ....
90. summarize—recapitulate ....
91. intrinsic—real ....
92. innocuous—harmful ....
93. exigency—emergency ....
94. tantamount—urgent ....
95. temerity—rashness ....

DIRECTIONS: Below are 10 groups of paired words, numbered 96 through 105. For each group, select the one pair of words, lettered (A), (B), (C), (D), (E), which are the opposite in meaning and mark your answer under the letter opposite that group.

Which of the following pair of words are opposite in meaning?

96. (A) wax—wane
(B) comfort—console
(C) exalt—rejoice
(D) enraged—angry
(E) tiny—minute

97. (A) becoming—fit
(B) commodious—roomy
(C) lack—plenty
(D) beg—supplicate
(E) compromise—conciliation

98. (A) certify—attest
(B) forbid—sanction
(C) kingly—regal
(D) quell—subdue
(E) quick—hasty

99. (A) concede—assent
(B) contemn—spurn
(C) convey—carry
(D) lag—hasten
(E) contrast—differentiate

100. (A) defer—postpone
(B) noble—base
(C) exigency—emergency
(D) fiery—ardent
(E) fault—flaw

101. (A) formidable—invincible
(B) futile—unavailing
(C) genuine—real
(D) halt—stop
(E) recant—admit

102. (A) alleviate—moderate
(B) apathy—unconcern
(C) turbid—clear
(D) arraign—charge
(E) austere—severe

103. (A) robust—unhealthy
(B) impediment—obstruction
(C) insidious—cunning
(D) novice—beginner
(E) obscure—hidden

104. (A) obstinate—stubborn
(B) pique—fret
(C) raze—destroy
(D) relinquish—renounce
(E) sinister—good

105. (A) ruthless—merciless
(B) support—sustain
(C) valid—sound
(D) timid—daring
(E) zeal—ardor

DIRECTIONS: Below are 10 groups of paired words, numbered 106 through 115. For each group, select the one pair of words, lettered (A), (B), (C), (D), (E), which are the same in meaning.

Which of the following pair of words are the same in meaning?

106. (A) transitory—permanent
(B) prohibit—allow
(C) beautiful—ugly
(D) broken—disunited
(E) ferocious—mild

107. (A) elucidate—clarify
(B) recent—ancient
(C) enthusiasm—apathy
(D) equivocal—indubitable
(E) evade—acknowledge

108. (A) extricate—imprison
(B) concur—endorse
(C) intimidate—assure
(D) lucid—obscure
(E) molest—comfort

109. (A) abandon—hold
(B) awkward—skillful
(C) consistent—varying
(D) constrain—beseech
(E) tedious—tiresome

110. (A) deficient—ample
(B) waste—conserve
(C) compromise—quarrel
(D) sanguine—optimistic
(E) desist—persevere

111. (A) monotony—variety
(B) remote—near
(C) propitiate—appease
(D) many—few
(E) veracity—deception

112. (A) fraud—honesty
(B) important—significant
(C) mollify—vex
(D) abate—maintain
(E) authorize—forbid

113. (A) eradicate—destroy
(B) barbarous—humane
(C) compulsion—freedom
(D) concur—differ
(E) incite—quell

114. (A) morose—cheerful
(B) munificient—penurious
(C) censorious—fault-finding
(D) predominate—subordinate
(E) extricate—bind

115. (A) gratify—displease
(B) grudge—good will
(C) interpose—withdraw
(D) irresponsible—accountable
(E) augment—increase

116. "A statement of self-evident truth" is called a (an)
(A) adage (B) axiom
(C) hypothesis (D) theory

117. "The path followed by one heavenly body in its revolution about another is called a (an)
(A) comet (B) meteor
(C) orbit (D) satellite

Each of the following items (numbered 118 to 130) contains one word which is not in keeping with the meaning which the sentence is evidently intended to carry. One of the five italicized words does most to spoil the true meaning of the item. In the space provided circle the letter which is below the incorrect word.

118. Many basements have old newspapers, excelsior, rags, and other highly *inflammable* material, while backyards may contain the same and other fire *hazards. Inspection* by the fire department and insistence that premises be placed in a less *permanent* condition doubtless do much to *reduce* the number of fires.
(A) inflammable (B) hazards
(C) inspection (D) permanent
(E) reduce

119. The Fire Department is *consigning* an *area comprising* 318 square *miles* with 578 miles of waterfront. Over 700,000 buildings with a population of 7,500,000 people cover this city to create one of the largest fire *protection* jobs in the world.
(A) consigning (B) area
(C) comprising (D) miles
(E) protection

120. It has been the belief of many persons that a well organized system for *educating* the *personnel* of the Fire Department and for enlightening the *public* on the *smoke* of fire hazards and fire protection has helped, in *no* small measure to reduce fire losses.
(A) educating (B) personnel
(C) public (D) smoke
(E) no

121. After a fire breaks out, there is almost bound to be *damage;* even where the *damage* itself is quickly extinguished, the *water* poured into a building may cause great destruction. Moreover, *firemen* cut holes in roofs and otherwise *mutilate* a property in order to get at the fire itself.
(A) damage (B) damage
(C) water (D) firemen
(E) mutilate

122. According to the statements of the National Board of Fire Underwriters, something like one fourth of all the fire loss in the United States is strictly *preventable,* and another third is classified as *due* to unknown causes. A great deal of this vague category probably involves carelessness and *consequently* might properly be *added* to the strictly *hazardous* class.
(A) preventable (B) due
(C) consequently (D) added
(E) hazardous

123. During every *ten* seconds more than one hundred dollars worth of property is destroyed by *gases* somewhere in the United States. Every hour of the day, and every day of the year, *fire* is taking its *fearful toll.*
(A) ten (B) gases
(C) fire (D) fearful
(E) toll

124. In recent years fire-fighting *apparatus* has been greatly improved. Almost everywhere motor-driven *pumping* engines have replaced steam *pipes,* and the *ladder* trucks of the larger cities are now *equipped* with quick-lifting aerial ladders.
(A) apparatus (B) pumping
(C) pipes (D) ladder
(E) equipped

125. With the development of modern *cities* and increasing *accretion,* not only in *business* districts but also in *residential* sections, the problems of fighting fire, together with its *prevention,* are becoming more important each year.
(A) cities (B) accretion
(C) business (D) residential
(E) prevention

126. Significant development in urban *fire* fighting did not take place until shortly after the middle of the nineteenth century, with the *advent* of the steam pumper *and* professionalized fire departments. Fire losses continued to *mount* for a long time after that date, but they must be *destroyed* by the changing values of destructible property.
(A) fire (B) advent
(C) and (D) mount
(E) destroyed

127. The costs of fire departments in cities of the United States mount high, as may be *hoped* from the very elaborate equipment and *extensive* staffs. On a per capita basis, *expenditures* for fire departments rank fifth among *costs* of operating and maintaining the general departments in *cities* of over 100,000 population.
(A) hoped (B) extensive
(C) expenditures (D) costs
(E) cities

128. *Sabotage* can be placed in three *introductions:* that of wrecking *machinery* of various types, *that* of wrecking human beings, and that of *wrecking* property by fire.
(A) sabotage (B) introductions
(C) machinery (D) that
(E) wrecking

129. Fire losses are often given on a per capita *basis,* which seems somewhat far-fetched, *since fires* are neither the unit *destroyed* nor are they considered as per capita *possessors* of fire-consumable wealth.
(A) basis (B) since
(C) fires (D) destroyed
(E) possessors

130. *Lack* of fire extinguishing systems and *lack* of ventilation have been a *device* to the Fire Department in that men attempting to enter cellars and subcellars through the *interior* stairways have been overcome by smoke, which has resulted in reduction of manpower at fires.
(A) lack (B) systems
(C) lack (D) device
(E) interior

Correct Answers For The Foregoing Questions

(Please make every effort to answer the questions on your own before looking at these answers. You'll make faster progress by following this rule.)

Vocabulary

1. B	18. A	34. C	50. B	66. A	82. C	98. B	114. C
2. A	19. B	35. B	51. A	67. C	83. B	99. D	115. E
3. B	20. E	36. B	52. D	68. D	84. E	100. B	116. B
4. E	21. D	37. D	53. A	69. A	85. C	101. E	117. C
5. C	22. C	38. D	54. C	70. B	86. O	102. C	118. D
6. C	23. B	39. D	55. A	71. B	87. S	103. A	119. A
7. D	24. D	40. E	56. A	72. B	88. O	104. E	120. D
8. D	25. C	41. A	57. B	73. C	89. N	105. D	121. B
9. C	26. C	42. B	58. E	74. A	90. S	106. D	122. E
10. B	27. C	43. B	59. C	75. C	91. S	107. A	123. B
11. A	28. E	44. D	60. A	76. D	92. O	108. B	124. C
12. D	29. C	45. D	61. D	77. A	93. S	109. E	125. B
13. E	30. B	46. D	62. B	78. A	94. N	110. D	126. E
14. A	31. E	47. E	63. A	79. C	95. S	111. C	127. A
15. C	32. A	48. B	64. B	80. A	96. A	112. B	128. B
16. C	33. A	49. C	65. D	81. B	97. C	113. A	129. E
17. C							130. D

SYNONYMS TEST ONE

DIRECTIONS: For each of the following questions, select the choice which best answers the question or completes the statement.

Correct key answers to all these test questions will be found at the end of the test.

1. An amendment is a
   (A) civic center (B) charter
   (C) penalty (D) change.

2. A quorum is a
   (A) minority (B) committee
   (C) majority (D) bicameral system.

3. Clearance refers to a
   (A) weight (B) hoistway
   (C) distance (D) cleaning process.

4. Pasteurized milk is milk that has been
   (A) watered (B) condemned
   (C) embargoed (D) purified.

5. Antitoxin is used in cases of
   (A) corrupt governmental officials
   (B) sanitary inspection
   (C) disease
   (D) elevator construction.

6. Libel refers to the
   (A) process of incurring financial liability
   (B) publication of a false statement which injures others
   (C) deportation of aliens
   (D) necessity for compulsory schooling.

7. Naturalization refers to the process of
   (A) becoming a civil service employee
   (B) being summoned to court
   (C) becoming a citizen
   (D) pledging allegiance to the American flag.

8. To comply with a rule means to
   (A) abide by a rule (B) abrogate a rule
   (C) dislike a rule (D) ignore a rule.

9. A budget is a
   (A) financial statement
   (B) method for training operators
   (C) device for insuring courtesy
   (D) means for selecting judges

10. A fulcrum is part of a
    (A) typewriter (B) lever
    (C) radio (D) lamp.

Answer Sheet

	A	B	C	D	E
1					
2					
3					
4					
5					
6					
7					
8					
9					
10					

SCORE

..................%

NO. CORRECT ÷ NO. OF QUESTIONS ON THIS TEST

Correct Answers

	A	B	C	D	E
1				▮	
2			▮		
3			▮		
4				▮	
5			▮		
6		▮			
7			▮		
8	▮				
9	▮				
10		▮			

SYNONYMS TEST TWO

DIRECTIONS: Each of the questions in this test is numbered and consists of pairs of words. Each pair of words is preceded by a letter. For each question, select the one pair of words wherein both words are synonymous. Mark the letter, A, B, C, D, or E, preceding that pair on your answer sheet.

Correct key answers to all these test questions will be found at the end of the test.

1. (A) transitory-permanent (B) prohibit-allow (C) beautiful-ugly (D) broken-disunited (E) ferocious-mild.

2. (A) elucidate-clarify (B) recent-ancient (C) enthusiasm-apathy (D) equivocal-indubitable (E) evade-acknowledge.

3. (A) extricate-imprison (B) concur-endorse (C) intimidate-assure (D) lucid-obscure (E) molest-comfort.

4. (A) abandon-hold (B) awkward-skillful (C) consistent-varying (D) constrain-beseech (E) tedious-tiresome.

5. (A) deficient-ample (B) waste-conserve (C) compromise-quarrel (D) sanguine-optimistic (E) desist-persevere.

6. (A) monotony-variety (B) remote-near (C) propitiate-appease (D) many-few (E) veracity-deception.

7. (A) fraud-honesty (B) important-significant (C) mollify-vex (D) abate-maintain (E) authorize-forbid.

8. (A) eradicate-destroy (B) barbarous-humane (C) compulsion-freedom (D) concur-differ (E) incite-quell.

9. (A) morose-cheerful (B) munificent-penurious (C) censorious-fault-finding (D) predominate-subordinate (E) extricate-bind.

10. (A) gratify-displease (B) grudge-good will (C) interpose-withdraw (D) irresponsible-accountable (E) augment-increase.

Correct Answers For The Foregoing Questions

To assist you in scoring yourself we have provided Correct Answers alongside your Answer Sheet. May we therefore suggest that while you are doing the test you cover the Correct Answers with a sheet of white paper.....to avoid temptation and to arrive at an accurate estimate of your ability and progress.

Answer Sheet

	A	B	C	D	E
1	⁞	⁞	⁞	⁞	⁞
2	⁞	⁞	⁞	⁞	⁞
3	⁞	⁞	⁞	⁞	⁞
4	⁞	⁞	⁞	⁞	⁞
5	⁞	⁞	⁞	⁞	⁞

Answer Sheet

	A	B	C	D	E
6	⁞	⁞	⁞	⁞	⁞
7	⁞	⁞	⁞	⁞	⁞
8	⁞	⁞	⁞	⁞	⁞
9	⁞	⁞	⁞	⁞	⁞
10	⁞	⁞	⁞	⁞	⁞

SCORE

................%

NO. CORRECT ÷
NO. OF QUESTIONS ON THIS TEST

Correct Answers

	A	B	C	D	E
1	⁞	⁞	⁞	▌	⁞
2	▌	⁞	⁞	⁞	⁞
3	⁞	▌	⁞	⁞	⁞
4	⁞	⁞	⁞	⁞	▌
5	⁞	⁞	⁞	▌	⁞

Correct Answers

	A	B	C	D	E
6	⁞	⁞	▌	⁞	⁞
7	⁞	▌	⁞	⁞	⁞
8	▌	⁞	⁞	⁞	⁞
9	⁞	⁞	▌	⁞	⁞
10	⁞	⁞	⁞	⁞	▌

SYNONYMS TEST THREE

Although vocabulary questions may not actually appear on your test. it is advisable to try to improve your vocabulary. Words and their meanings are quite important on reading, writing and correct usage tests. By broadening your vocabulary, you will definitely improve your marks in these and similar subjects.

DIRECTIONS: For each of the following statements, mark T if the statement is True, and F if it is False.

1. A *competent* employee is one who is slow and inefficient. T F
2. A person who commits *perjury* does not tell the truth. T F
3. The *prosecutor* in a criminal case is the lawyer who presents evidence against the defendant. T F
4. A *destitute* person has a large amount of money. T F
5. A person with a *florid* complexion has a pale face. T F
6. A noise that is *audible* is capable of being heard. T F
7. Anyone who is *agile* is quick and nimble. T F
8. An employee who gives information in a *curt* manner is sympathetic and courteous. T F
9. A person who is *prudent* is careless in his attention to duty. T F
10. If a person pays an *exorbitant* amount of money for an article, he is paying a fair price. T F

Now, push forward! Test yourself and practice for your test with the carefully constructed quizzes that follow. Each one presents the kind of question you may expect on your test. And each question is at just the level of difficulty that may be expected. Don't try to take all the tests at one time. Rather, schedule yourself so that you take a few at each session, and spend approximately the same time on them at each session. Score yourself honestly, and date each test. You should be able to detect improvement in your performance on successive sessions.

A portion of the standard answer sheet is provided after each test for marking your answers in the way you'll be required to do on the actual exam. At the right of this answer sheet, to make the scoring job simpler (after you have derived your own answers), you'll find our correct answers.

SCORE

..................%

NO. CORRECT ÷

NO. OF QUESTIONS ON THIS TEST

Correct Answers

(You'll learn more by writing your own answers before comparing them with these.)

1. F	6. T
2. T	7. T
3. T	8. F
4. F	9. F
5. F	10. F

SYNONYMS TEST FOUR

DIRECTIONS: This test contains a list of numbered words, and other words divided into five groups—Group A, Group B, Group C, Group D and Group E. For each of the numbered words, select that word in one of the five groups which is most nearly the same in meaning. The letter of that group is the answer for that item. Your answer sheet and our correct answers follow the last question.

Correct key answers to all these test questions will be found at the end of the test.

Group A

indication ambiguous excruciating thin

Group B

confirmation financial phobia erasure

Group C

fiduciary similar yellowish skill

Group D

theft winding receive procrastination

Group E

franchise heir hardy preference

1. fiscal
2. deletion
3. equivocal
4. corroboration
5. tortuous
6. predilection
7. sallow
8. virtuosity
9. scion
10. tenuous

Correct Answers For The Foregoing Questions

To assist you in scoring yourself we have provided Correct Answers alongside your Answer Sheet. May we therefore suggest that while you are doing the test you cover the Correct Answers with a sheet of white paper.....to avoid temptation and to arrive at an accurate estimate of your ability and progress.

Answer Sheet

	A	B	C	D	E
1					
2					
3					
4					
5					
6					
7					
8					
9					
10					

SCORE

..................%

NO. CORRECT ÷ NO. OF QUESTIONS ON THIS TEST

Correct Answers

	A	B	C	D	E
1		▮			
2		▮			
3	▮				
4		▮			
5				▮	
6					▮
7			▮		
8			▮		
9					▮
10	▮				

SYNONYMS TEST FIVE

DIRECTIONS: This test contains a list of numbered words, and other words divided into five groups—Group A, Group B, Group C, Group D and Group E. For each of the numbered words, select that word in one of the five groups which is most nearly the same in meaning. The letter of that group is the answer for that item. Your answer sheet and our correct answers follow the last question.

Correct key answers to all these test questions will be found at the end of the test.

1. prophylactic
2. palliation
3. redolent
4. indictment
5. misfeasance
6. holograph
7. ancillary
8. hectic
9. obsolescence
10. holocaust

Group A

ruddy retribution mitigation decadence

Group B

dental fragrant accusation symptom

Group C

preventive destruction aggravation testimony

Group D

intimidation subsidiary feverish trespass

Group E

consecration excited corpulence handwritten

Correct Answers For The Foregoing Questions

To assist you in scoring yourself we have provided Correct Answers alongside your Answer Sheet. May we therefore suggest that while you are doing the test you cover the Correct Answers with a sheet of white paper.....to avoid temptation and to arrive at an accurate estimate of your ability and progress.

Answer Sheet

	A	B	C	D	E
1					
2					
3					
4					
5					
6					
7					
8					
9					
10					

SCORE

.................%

NO. CORRECT ÷
NO. OF QUESTIONS ON THIS TEST

Correct Answers

	A	B	C	D	E
1			▮		
2	▮				
3		▮			
4		▮			
5				▮	
6					▮
7				▮	
8				▮	
9	▮				
10			▮		

SYNONYMS TEST SIX

DIRECTIONS: For each question in this test, select the appropriate letter preceding the word which is most nearly the same in meaning as the italicized word in each sentence.

Correct key answers to all these test questions will be found at the end of the test.

1. The person who is *diplomatic* in his relations with others is, most nearly
(A) well dressed (B) very tactful
(C) somewhat domineering (D) deceitful and tricky
(E) verbose.

2. Action at this time would be *inopportune*. The word "inopportune" means most nearly
(A) untimely (B) premeditated
(C) sporadic (D) commendable
(E) fortunate.

3. The word *appraise* means most nearly
(A) consult (B) attribute
(C) manage (D) honor
(E) judge.

4. The word *cognizant* means most nearly
(A) rare (B) reluctant
(C) aware (D) haphazard
(E) correlated.

5. *Probity* is an important requirement of many positions. The word "probity" means most nearly
(A) analytical ability (B) vision
(C) tried integrity (D) clear insight
(E) perseverence.

6. The word *denote* means most nearly
(A) encumber (B) evade
(C) furnish (D) indicate
(E) reduce in rank.

7. The competent employee should know that a method of procedure which is *expedient* is most nearly
(A) unchangeable (B) based upon a false assumption
(C) unduly harmful (D) difficult to work out
(E) suitable to the end in view.

8. An incentive which is *potent* is most nearly
(A) impossible (B) highly effective
(C) not immediately practicable (D) a remote possibility
(E) universally applicable.

9. An employer who is *judicious* is most nearly
(A) domineering (B) argumentative
(C) sincere (D) arbitrary
(E) wise.

10. He presented a *controversial* plan. The word "controversial" means most nearly
(A) subject to debate (B) unreasonable
(C) complex (D) comparable
(E) well formulated.

Answer Sheet

	A	B	C	D	E
1	⁞	⁞	⁞	⁞	⁞
2	⁞	⁞	⁞	⁞	⁞
3	⁞	⁞	⁞	⁞	⁞
4	⁞	⁞	⁞	⁞	⁞
5	⁞	⁞	⁞	⁞	⁞
6	⁞	⁞	⁞	⁞	⁞
7	⁞	⁞	⁞	⁞	⁞
8	⁞	⁞	⁞	⁞	⁞
9	⁞	⁞	⁞	⁞	⁞
10	⁞	⁞	⁞	⁞	⁞

SCORE

.... %

NO. CORRECT ÷ NO. OF QUESTIONS ON THIS TEST

Correct Answers

	A	B	C	D	E
1		■			
2	■				
3					■
4			■		
5			■		
6				■	
7					■
8		■			
9					■
10	■				

SYNONYMS TEST SEVEN

DIRECTIONS: For each question in this test, select the appropriate letter preceding the word which is most nearly the same in meaning as the italicized word in each sentence.

Correct key answers to all these test questions will be found at the end of the test.

1. "He sent the irate employee to the personnel manager." The word *irate* means most nearly
   (A) irresponsible (B) untidy
   (C) insubordinate (D) angry.
2. An *ambiguous* statement is one which is
   (A) forceful and convincing
   (B) capable of being understood in more than one sense
   (C) based upon good judgment and sound reasoning processes
   (D) uninteresting and too lengthy.
3. To *extol* means most nearly to
   (A) summon (B) praise
   (C) reject (D) withdraw.
4. The word *proximity* means most nearly
   (A) similarity (B) exactness
   (C) harmony (D) nearness.
5. "His friends had a detrimental influence on him." The word *detrimental* means most nearly
   (A) favorable (B) lasting
   (C) harmful (D) short-lived.
6. "The chief inspector relied upon the veracity of his inspectors." The word *veracity* means most nearly
   (A) speed (B) assistance
   (C) shrewdness (D) truthfulness.
7. "There was much diversity in the suggestions submitted." The word *diversity* means most nearly
   (A) similarity (B) value
   (C) triviality (D) variety.
8. "The survey was concerned with the problem of indigence." The word *indigence* means most nearly
   (A) poverty (B) corruption
   (C) intolerance (D) morale.
9. "The investigator considered this evidence to be extraneous." The word *extraneous* means most nearly
   (A) significant (B) pertinent but unobtainable
   (C) not essential (D) inadequate.
10. "He was surprised at the temerity of the new employee." The word *temerity* means most nearly
    (A) shyness (B) enthusiasm
    (C) rashness (D) self-control.

Correct Answers

Answer Sheet

	A	B	C	D	E
1					
2					
3					
4					
5					
6					
7					
8					
9					
10					

SCORE

.................%

NO. CORRECT ÷ NO. OF QUESTIONS ON THIS TEST

Correct Answers

	A	B	C	D	E
1				■	
2		■			
3		■			
4				■	
5			■		
6				■	
7				■	
8	■				
9			■		
10			■		

SYNONYMS TEST EIGHT

DIRECTIONS: In each of the sentences below, one word is in italics. Following each sentence are four or five lettered words or phrases. For each sentence, choose the letter preceding the word or phrase which most nearly corresponds in meaning with the italicized word.

Correct key answers to all these test questions will be found at the end of the test.

1. The change in procedure *stimulated* the men
   (A) rewarded (B) antagonized
   (C) gave an incentive to
   (D) restricted the activities of
   (E) lowered the efficiency of.
2. Courage is a trait difficult to *instill*. The word "instill" means most nearly
   (A) measure exactly (B) impart gradually
   (C) predict accurately (D) restrain effectively
   (E) discuss meaningfully.
3. The vehicle was left *intact*. The word "intact" means most nearly
   (A) a total loss (B) unattended
   (C) where it could be noticed (D) undamaged
   (E) repaired.
4. The witness was *recalcitrant*. The word "recalcitrant" means most nearly
   (A) cooperative (B) delirious
   (C) highly excited
   (D) accustomed to hard work
   (E) stubbornly resistant.
5. A *conscientious* person is one who
   (A) feels obligated to do what he believes right
   (B) rarely makes errors
   (C) frequently makes suggestions for procedural improvements
   (D) has good personal relationships with others
   (E) is consistent in his behavior.
6. It was reported that *noxious* fumes were escaping. The word "noxious" means most nearly
   (A) concentrated (B) gaseous
   (C) greenish colored (D) heavy
   (E) harmful.
7. A person with a *sallow* complexion was seen near the scene. The word "sallow" means most nearly
   (A) ruddy (B) dark
   (C) pale and yellowish (D) highly freckled
   (E) red and florid.
8. The word *cogent* means most nearly
   (A) confused (B) opposite
   (C) unintentional (D) convincing
   (E) irrelevant.
9. The word *divergent* means most nearly
   (A) simultaneous (B) differing
   (C) approaching (D) parallel
   (E) twisting.
10. The word *ostensibly* means most nearly
   (A) undoubtedly (B) infrequently
   (C) powerfully (D) apparently
   (E) slowly.

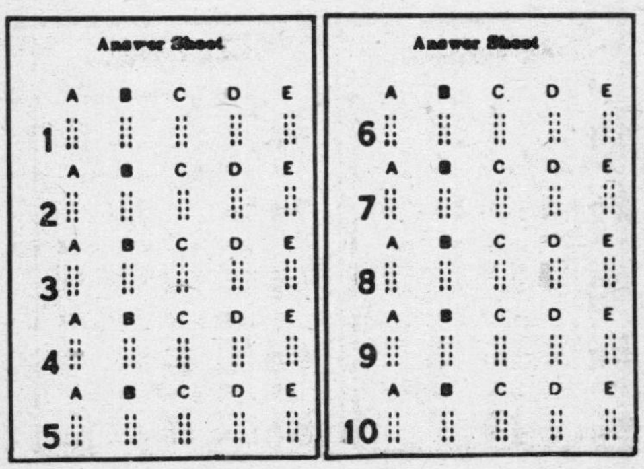

SCORE

..................%

NO. CORRECT ÷

NO. OF QUESTIONS ON THIS TEST

Correct Answers

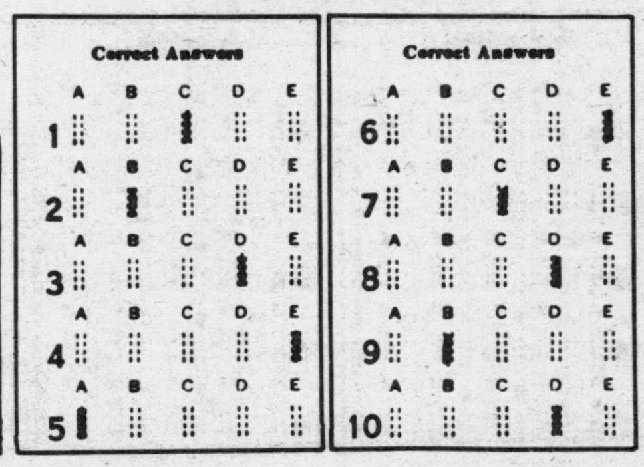

SYNONYMS TEST NINE

DIRECTIONS: In each of the sentences below, one word is in italics. Following each sentence are four or five lettered words or phrases. For each sentence, choose the letter preceding the word or phrase which most nearly corresponds in meaning with the italicized word.

Correct key answers to all these test questions will be found at the end of the test.

1. To say that the work is *tedious* means, most nearly, that it is
   (A) technical (B) interesting
   (C) tiresome (D) confidential.
2. A *vivacious* person is one who is
   (A) kind (B) talkative
   (C) lively (D) well-dressed.
3. An *innocuous* statement is one which is
   (A) forceful (B) harmless
   (C) offensive (D) brief.
4. To say that the order was *rescinded* means, most nearly, that the order was
   (A) revised (B) canceled
   (C) misinterpreted (D) confirmed.
5. To say that the administrator *amplified* his remarks means, most nearly, that the remarks were
   (A) shouted (B) expanded
   (C) carefully analyzed (D) summarized briefly.
6. "Peremptory commands will be resented in any organization." The word *peremptory* means most nearly
   (A) unexpected (B) unreasonable
   (C) military (D) dictatorial.
7. A person should know the word *sporadic* means, most nearly
   (A) occurring regularly (B) sudden
   (C) scattered (D) disturbing.
8. To *oscillate* means, most nearly, to
   (A) lubricate (B) waver
   (C) decide (D) investigate.
9. A *homogeneous* group of persons is characterized by its
   (A) similarity (B) teamwork
   (C) discontent (D) differences.
10. A *vindictive* person is one who is
    (A) prejudiced (B) unpopular
    (C) petty (D) revengeful.

Correct Answers

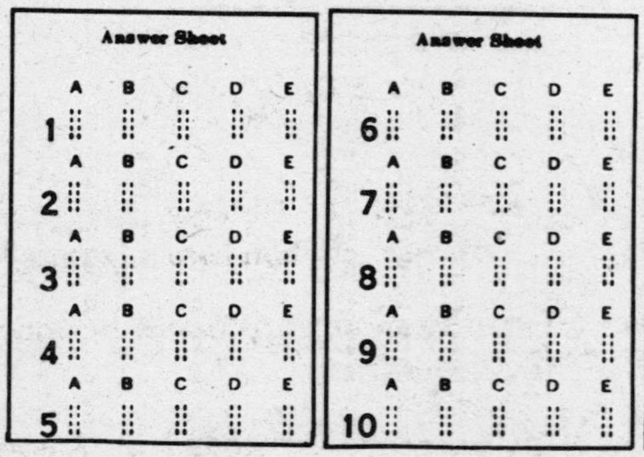

SCORE

..................%

NO. CORRECT ÷
NO. OF QUESTIONS ON THIS TEST

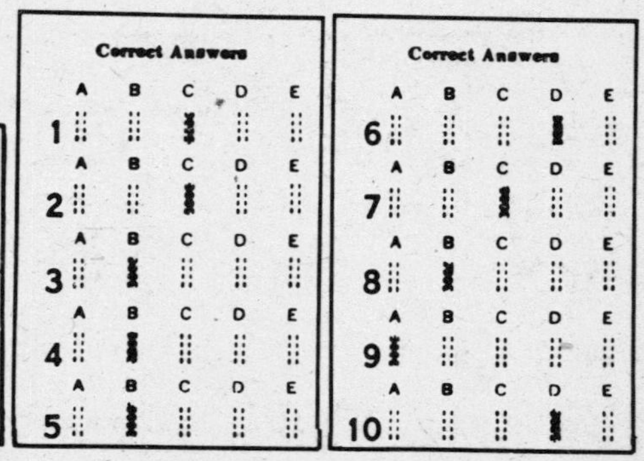

NUMERICAL RELATIONS

Solutions To Basic Problems

Profit and Loss

1. Hammers are bought for $18.00 a dozen. In order to gain 40%, what must the selling price per hammer be?

(A) $2.10 (B) $2.00
(C) $2.50 (D) $3.00

SOLUTION: To find the selling price we must multiply the cost by the rate of profit or loss.

In this case the cost is $18.00 divided by 12; $18.00 being the cost of an entire dozen. Since there is a profit of 40% we must multiply the cost by 1.40 or 140%. (If we sold something at 100% of its cost, we should be getting what we paid for it.)

$$\frac{18}{12} \times 1.40 = \begin{array}{r} 18 \\ \underline{\times\ 1.40} \\ 720 \\ \underline{18} \\ 25.20 \end{array}$$

$$12\,\overline{)\,25.20}\quad 2.10 \qquad \textbf{Answer}$$

2. Hammers are bought for $30.00 a dozen and sold at $3.50 each. The rate of profit on the transaction is:

(A) 30% (B) 40%
(C) 50% (D) 45%

SOLUTION: To find the rate of profit or loss, we first find the actual profit or loss and then find what percent of the COST this is.

Multiplying $3.50 by 12 to find the cost of a dozen hammers:

$$\$3.50 \times 12 = \$42.00$$

Subtracting the cost from the selling price to find the actual profit:

$$\begin{array}{r} \$42.00 \\ -\ \underline{30.00} \\ \$12.00 \end{array}$$

Finding what percent of the cost the profit is involves converting a fraction into a percent.

We multiply the fraction by 100 and perform the indicated divisions:

$$\frac{\$12}{\$30} \times 100 = \quad 3\cancel{0}\,\overline{)\,120\cancel{0}}\quad 40\% \qquad \textbf{Answer}$$

Addition of Fractions

3. If we add 8-1/5, 45-5/8, 2-17/20, 14-1/2, and 1-21/40 the answer will be
(A) 70-8/10 (B) 72-7/10
(C) 72-1/5 (D) 70-7/10.

SOLUTION: In adding mixed numbers like these, we perform three additions: The addition of the whole numbers, the addition of the fractions, and combination of the added whole numbers and fractions. Adding the whole numbers presents no difficulty. Our sum is 70.

$$8 + 45 + 2 + 14 + 1 = 70$$

To add the fractions we must first find the least common denominator. In this case it is 40.

Then for each separate fraction we divide the denominator into the common denominator and multiply the resulting quotient by the numerator.

We add all these products and divide by the common denominator. Here are the actual calculations:

	40
8-1/5	8
45-5/8	25
2-17/20	34
14-1/2	20
1-21/40	21

$$\frac{108}{40} = 2\frac{28}{40} = 2\frac{7}{10}$$

$$\begin{array}{r} 70 \\ +\ 2\frac{7}{10} \\ \hline 72\frac{7}{10} \end{array}$$

Answer

Interest

4. $1,850 is invested for 50 days at a rate of 5%. The interest return is
(A) $12.00 (B) $10.00
(C) $12.67 (D) $13.00.

SOLUTION: While there are many short cuts used by banks and commercial houses in computing interest, the best plan for the candidate is to understand thoroughly all the steps involved in this computation and to use them all intelligently.

Expressed as a formula, interest may be computed thus: P × R × T = I. If we are given interest, principal and time, and asked to find the Rate, the operation may be expressed thus:

$$R = \frac{I}{P \times T}$$

And if we are given R, T, and I, and asked to find the Principal, this is the formula:

$$P = \frac{I}{R \times T}$$

If we multiply the principal by the rate of interest, and the length of time the money draws interest, we have the amount of interest due.

If the money were to bear interest for a year we would only have to multiply $1,850 by 5% (5/100) in this example. However, the money only bears interest for 50 days or 50/365 of a year. The rest is done by simple cancellation.

$$\frac{\overset{\$925}{\$1850} \times \overset{1}{5} \times \overset{1}{\cancel{50}}}{\underset{\underset{1}{\cancel{2}}}{\cancel{100}}\ \ \underset{73}{\cancel{365}}} = \frac{925}{73} = \$12.67$$ **Answer**

5. What amount of money yields $40.00 per month if invested at an annual rate of 5%
(A) $9,000 (B) $9,500
(C) $9,600 (D) $9,400.

SOLUTION: To find the principal we must divide interest by rate by time.

If we multiply $40.00 per month by twelve we find the interest yield for a year. $40 × 12 = $480.00. The Time factor in this problem is now one year.

For the rest we have only to follow out our formula:

$$P = \frac{I}{R \times T} = P = \frac{\$480}{\frac{5}{100} \times 1}$$

DIVISION OF FRACTIONS: At this point an interesting difficulty presents itself: The division of fractions. To divide $80 by 5/100 we have simply to invert the fraction and multiply thus:

$$\$480 \div \frac{5}{100} = \$\overset{96}{\cancel{480}} \times \frac{100}{\underset{1}{\cancel{5}}} = \$9{,}600 \quad \textbf{Answer}$$

6. A woman invested $4,000 in a speculative venture for 9 months. A second woman invested $6,000 in the same business for 6 months. The net gain was $720.00. What was the second woman's return on her investment if all the profits were divided between the two women?
(A) $360 (B) $700
(C) $400 (D) $350.

SOLUTION: So far as regards dividends, the investment of $6,000 for 6 months is the same as the investment of $36,000 for 1 month. The same statement can be made regarding $4,000 for 9 months.

First Woman — $4,000 for 9 months = $36,000 for 1 month.

Second Woman— $6,000 for 6 months = $36,000 for 1 month.

Both $72,000 for 1 month.

It is a coincidence that the shares of the two women in this example are the same. However, the procedure here is exactly the same as though they had different shares.

We find what part the second woman's share bears to the total. We then take this proportion of the total income and our result is the second woman's share.

$$\frac{\overset{1}{\cancel{36{,}000}}}{\underset{2}{\cancel{72{,}000}}} \times \overset{360}{\cancel{720}} = \$360 \quad \textbf{Answer}$$

7. The rate of interest on a principal of $10,000 that will yield $80.00 in 65 days is
(A) 4% (B) 5.5%
(C) 4.49% (D) 6%.

SOLUTION: The formula here is

$$R = \frac{I}{P \times T}$$

Substituting, we have:

$$R = \frac{\$80}{10{,}000 \times \frac{65}{365}} = \frac{\$80}{\frac{650{,}000}{365}} =$$

$$\$80 \times \frac{365}{650{,}000} = \frac{29{,}200}{650{,}000}$$

CONVERTING A FRACTION INTO A PERCENT: This is a simple operation if the proper steps be known and taken. We must convert the fraction $\frac{29{,}200}{650{,}000}$ into a percent so that we may properly express the rate of interest.

To change a fraction into a percent we must multiply the fraction by 100 and then carry through the indicated division.

$$\frac{29{,}200}{650{,}000} \times 100 = \frac{2{,}920{,}000}{650{,}000} =$$

$$65\cancel{0{,}000}\,\overline{)\,2{,}92\cancel{0{,}000}} = 4.49\% \quad \textbf{Answer}$$

Assessment

8. If a piece of property is assessed at \$45,700 and the tax rate on real property is \$2.40 per \$1,000, the amount of tax that must be paid on this property is

(A) \$110
(B) \$112
(C) \$109.68
(D) \$109.

SOLUTION: Since the tax rate is \$2.40 per \$1,000, we must determine how many thousands of dollars are involved. To do this, we divide \$45,700 by \$1,000. And the result is 45.7.

$$45{,}700 \div 1{,}000 = 45.7$$

When we multiply 45.7 by \$2.40, we learn the amount of the tax—

$$45.7 \times \$2.40 = \$109.68$$ **Answer**

9. \$60,000 worth of land is assessed at 120% of its value. If the tax rate is \$2.56 per 1,000 the amount of tax to be paid is

(A) \$190
(B) \$195
(C) \$184.32
(D) \$180.

SOLUTION: To find how much the land has been assessed:

$$60{,}000 \times \frac{12\not{0}}{10\not{0}} =$$

$$\frac{720{,}00\not{0}}{1\not{0}} = \$72{,}000$$

If tax is \$2.56 per \$1,000, multiply:

$$\begin{array}{r} 2.56 \\ \times\ 72 \\ \hline 512 \\ 1792 \\ \hline \$184.32 \end{array}$$ **Answer**

Cubic Volume

10. A bin measures 14 feet by 9 feet by 7½ feet. Allowing 4/5 bushel of grain per cubic foot, how many bushels will the bin hold?

SOLUTION:

Length × Width × Height = Cubic Area.

$$L \times W \times H = \text{Cubic Area}$$

$$14 \times 9 \times 7\tfrac{1}{2} = \overset{7}{\not{14}} \times 9 \times \frac{15}{\not{2}} = 945 \text{ cubic ft.}$$

Since each cubic foot of space holds 4/5 bushel of grain:

$$\overset{189}{\not{945}} \times \frac{4}{\not{5}} = 756 \text{ Bushels.}$$ **Answer**

Literal Problems

11. If L explosions occur during a given month and result in Q dollars of loss, the average loss per explosion in dollars is:

(A) $L \times Q$
(B) Q/L
(C) L/A
(D) $12K/2P$
(E) none.

SOLUTION: This is a simple problem in determining an average. The presence of letters rather than numbers makes it slightly more difficult by imposing upon us the burden of using fundamental principles rather than habitual modes of action.

The average loss per explosion is the total loss divided by the number of explosions, or the average loss per explosion equals

$$\frac{\text{Total amount of loss}}{\text{Number of explosions}}$$

Since the total loss in a given month is Q and the number of explosions is L we may say that the average loss per explosion is $\frac{Q}{L}$ **Answer**

12. If there is a total of J garbage trucks in operation in one city covering a total of E miles per hour, we can find the average street mileage per truck from the above data, without considering

(A) the number of cars
(B) the total street mileage
(C) the average speed
(D) any of these values
(E) any further data besides the above.

SOLUTION: The average street mileage per truck is the total street mileage divided by the total number of trucks, that is, the average street mileage per truck $= \frac{\text{Total street mileage}}{\text{Total No. of trucks}}$ (if we had 20 trucks covering a total street mileage of 300 miles then the average mileage covered by each truck is 15 miles or

$$\frac{300}{20} = 15 \text{ miles.)}$$

From this it is clear that the average speed of the trucks does not come into the consideration of the average mileage per truck and therefore (C) is correct. **Answer**

13. During July, T families took out insurance policies, representing an increase of M families over the number taking them in May. In June, however, the number taking out insurance was P less than in May. If there were R insurance agents in each of these 3 months, the average number of policies written per insurance agent in June was:

(A) $\frac{T - M}{P \quad R}$ (B) $\frac{T - M - P}{R}$
(C) $\frac{M + T - R}{R}$ (D) $\frac{T + M + P}{R}$

SOLUTION: We must first determine how many people took out insurance in June and get this quantity in terms of T, M, and P.
in July—T families took out insurance.
in May—T - M families took out insurance (since July is an increase of M over May).
in June—(no. in May) - P (since it was P less than May) = T - M - P - Total number of policies issued in May.

Now the average number of policies per insurance agent:

$$\frac{\text{Total number of policies in June}}{\text{Total number of insurance agents in June}} = \frac{T - M - P}{R}$$ **Answer**

14. Clerk A sorts B letters per hour, clerk C sorts D letters per hour. The D letters which clerk C sorts exceed those which clerk A sorts by 10 letters per hour. Measured in number sorted per 8-hour day, clerk C exceeds A by:

(A) $D - B \times 10$ (B) $(D - B) \times 8$
(C) $D + C - A + B$ (D) $C + D - A + B$

SOLUTION: The D letters sorted per hour by clerk C exceeds (is greater than) the B letters sorted per hour by clerk A, by an amount of 10 letters per hour, or D is 10 more than B. In symbols: $D = B + 10$.

In 8 hours, clerk C will have sorted 8 D letters while clerk A will have sorted 8 B letters and since the difference in one hour is 10 letters, in 8 hours, the difference will be 8×10. $8D = 8B + 80$ or bringing 8 B to the other side: $8D - 8B = 80$ or factoring out the 8 on the left.

We have $8(D - B) = 80$ which is the amount of letters by which clerk C's output exceeds clerk A's output and hence answer (B) above is correct. **Answer**

15. The annual salary of a machinist is R dollars more than that of his assistant. His assistant earns V dollars annually. The amount in monthly salary, by which the machinist exceeds his assistant is given by:

(A) $\frac{V - R}{13}$ (B) $\frac{RV}{12}$
(C) $R - V$ (D) $12V - R$
(E) $\frac{R}{12}$

SOLUTION: The annual salary of the machinist is $V + R$ dollars. The annual salary of his assistant is V dollars. The monthly salary of the machinist is $\frac{V}{12} + \frac{R}{12}$ dollars. (Since in a year the machinist receives 12 times as much as he receives in a month, we divide the yearly salary by 12 to find the amount earned in one month.) Similarly, the monthly salary of his assistants is $\frac{V}{12}$ dollars. Therefore the machinist's monthly salary exceeds his assistant's salary, monthly, by $\frac{R}{12}$ dollars.

Answer (E) **is correct.**

16. A family of 5 has two employed members earning L dollars a month. The family receives a total semi-monthly relief allowance of M dollars. If the rent allowance is N dollars, and the amount spent for food is twice that for rent, the amount spent monthly for all items other than food and rent is:

(A) $L + 2M - 3N$

(B) $N + L + M$

(C) $L + M - 2N$

SOLUTION: The amount spent monthly for all items other than food and rent is the total income for one month minus the total expenditure for food and rent.

Total income $= L + 2M$ (since M is a semi-monthly allowance, the monthly allowance is 2M or M multiplied by 2.)

Total exp. for food and rent $=$ N (for rent) $+ 2N$ (food)—since the amount spent for other items $= L + 2M - N - 2N =$
$L + 2M - 3N$. **Answer**

17. If psychological studies of college students show K per cent to be emotionally unstable, the number of college students not emotionally unstable per one hundred college students is:

(A) 100 minus K
(B) 100 times (K minus)
(C) K minus 1

SOLUTION: Since K percent $= \frac{K}{100}$ in 100 students, K% of 100 are emotionally unstable then $\frac{K}{100} \times 100 = K$ students are unstable.

Therefore the remaining students are emotionally stable and they number $100 - K$.

Answer (A) **is correct.**

Rate, Time, and Distance Problems

In all these problems the formula to be followed is very simple:

Rate (speed) X Time = Distance.

If you are given the RATE and DISTANCE and are asked to find time then you simply make the obvious modification in the formula:

$$\text{Time} = \frac{\text{Distance}}{\text{Rate}}$$

To find rate given distance and time:

$$\frac{\text{Distance}}{\text{Time}} = \text{Rate}$$

There are many complications that can be introduced but if these fundamental ideas can be kept clearly in view few difficulties will be encountered.

---

18. Two hikers start walking from the city line at different times. The second hiker whose speed is 4 miles per hour starts 2 hours after the first hiker whose speed is 3 miles per hour. Determine the amount of time and distance that will be consumed before the second hiker catches up with the first.

SOLUTION: Since the first man has a 2 hour headstart and is walking at the rate of 3 miles per hour he is 6 miles from the city line when the second hiker starts.

Rate $\times$ Time = Distance.

Subtracting 3 miles per hour from 4 miles per hour gives us 1 mile per hour or the difference in the rates of speed of the two men. In other words, the second hiker gains one mile on the first hiker in every hour.

Since there is a 6 mile difference to cut down and it is cut down one mile every hour, it is clear that the second hiker will need 6 hours to overtake his companion.

In this time he will have traveled $4 \times 6 = 24$ or 24 miles. The first hiker will have been walking 8 hours since he had a 2 hour headstart $8 \times 3 = 24$.

Time = 6 hours

Distance = 24 miles **Answer**

19. The same two hikers start walking toward each other along a road connecting two cities which are 60 miles apart. Their speeds are the same as in the preceding problem, 3 and 4 miles per hour. How much time will elapse before they meet?

SOLUTION: In each hour of travel toward each other the men will cut down a distance equal to the sum of their speeds. $3 + 4 = 7$ miles per hour. To meet they must cut down 60 miles, and at 7 miles per hour this would be

$$\frac{D}{R} = T \quad \frac{60}{7} = 8\frac{4}{7} \text{ hours.}$$ **Answer**

20. The problem might also have asked: "How much distance must the slower man cover before the two hikers meet?" In such case we should have gone through the same steps plus one additional step:

The time consumed before meeting was $8\frac{4}{7}$ hours. To find the distance covered by the slower hiker we merely multiply his rate by the time elapsed.

$$R \times T = D \quad 3 \times 8\frac{4}{7} = 25\frac{5}{7}$$ **Answer**

Time and Work Problems

21. If A does a job in 6 days, and B does the same job in 3 days, how long will it take the two of them, working together, to do the job?

SOLUTION: Almost any problem of this type can be solved quite simply by fractions, without resorting to higher mathematics.

A. If A does the whole job in 6 days, he will do 1/6 of the job in one day.
If B does the whole job in 3 days, he will do 1/3 of the job in one day.

B. 1/3 + 1/6 = 1/2

C. 1/2 of the job will be finished in one day if the two men work together.

D. The whole job will be finished in two days.

Answer

EXPLANATION OF SOLUTION:

A. If you are given the time that a job takes you have merely to find the reciprocal of that time in order to find how much of the work would be done in one day. Finding the reciprocal simply means inverting the figure.

If you do a job in 2 1/2 days, you would do 2/5 of the job in one day.

2 1/2 = 5/2

Finding the reciprocal or inverting:

5/2 = 2/5

Another way of looking at the same operation:

All of the work is done in 5/2 days. In other words it takes 5 half days to finish the job. In one half day 1/5 of the job would be completed, and consequently in one day (2/2) 2/5 of the job would be completed.

CAUTION: If the total time for the job is given in HOURS you will, by getting the reciprocal, find what fraction of the work is done in one HOUR. The procedure for the rest of the problem, of course, is the same as above, except that the answer is in hours.

B. The total time must be reduced to a fraction of the total job because it would not do to simply add the time consumed by each man. 3 days and 6 days added together yield nine days, which is merely TIME and tells us nothing of the AMOUNT OF WORK. But 1/3 and 1/6 do represent amounts of work.

C. Adding these two fractions together we discover the part of the job that would be completed in one day.

D. If we are told that a certain fraction represents the amount of work done in one day and if we wish to find how long the entire job would take, we find the reciprocal of the fraction.

1/2 in 1 day $\quad \frac{2}{2}$ (or all) in $\frac{2}{1}$ days.

Two principles should be kept in mind.

1. To find the part, invert the time.
2. To find the time, invert the part.

22. A and B working together do a job in $4\frac{1}{2}$ days. B, working alone, is able to do the job in 10 days. How long would it take A, working alone, to do the job?

SOLUTION: All of the job in 9/2 days. $\frac{2}{9}$ of the job in 1 day.
If B takes 10 days to do the job alone he will do 1/10 of the job in one day.

To find the work done by A in one day we subtract B's work from the amount of work done by the two men together in one day.

$$2/9 - 1/10 = \frac{20-9}{90} = \frac{11}{90}$$

11/90 represents the portion of the total job done by A in one day.

Inverting, we found how long it would take him to do the entire job.

$$\frac{90}{11} = 8\frac{2}{11} \text{ Days.}$$ **Answer**

23. If A can do a job in 6 days which B can do in 5-1/2 days, and C can do in 2-1/5 days, how long would the job take if A, B, and C were working together?

SOLUTION:

A Does the job in 6 days: 1/6 of the job in 1 day.
B Does the job in 5-1/2 days: 2/11 of the job in 1 day.
C Does the job in 2-1/5 days: 5/11 of the job in 1 day.

Adding the work done by A, B, and C in one day to find the work done by all three in one day.

$$\frac{1}{6} + \frac{2}{11} + \frac{5}{11} = \frac{11 + 12 + 30}{66} = \frac{53}{66}$$

Finding the reciprocal of $\frac{53}{66}$ in order to find how long the total job would take:

$$\frac{66}{53} = 1\frac{13}{53} \text{ Days.}$$ **Answer**

24. One pipe fills a pool in 20 minutes, a second can fill the pool in 30 minutes, and a third can fill it in 10 minutes. How long would it take the three together to fill the pool?

SOLUTION: First pipe—fills in 20 minutes—fills 1/20 of pool in 1 minute.
Second pipe—fills in 30 minutes—fills 1/30 of pool in 1 minute.
Third pipe—fills in 10 minutes—fills 1/10 of pool in 1 minute.

Adding three fractions together to determine what part of the pool will be filled in one minute when the three pipes are working together.

$$1/20 + 1/30 + 1/10 = \frac{3 + 2 + 6}{60} = \frac{11}{60}$$

If 11/60 of the pool is filled in one minute the reciprocal of the fraction will tell us how many minutes will be required to fill the whole pool.

$$\frac{60}{11} = 5\frac{5}{11} \text{ Minutes.}$$ **Answer**.

Problems in Proportions

25. If 5 men can build 6 miles of railroad track in 40 days, how many miles of track can be built by 3 men in 15 days?

SOLUTION: One of the best methods of solving such problems is by directly making the necessary cancellations, divisions, and multiplications.

In this example, we wish to find how many miles of track will be built if both the number of workers and the working time are reduced. It is easily seen that the amount of track constructed will be less than under the old conditions. But how much less?

Since we now have 3 men where before there were 5, we may assume that so far as manpower is concerned, production will be 3/5 as high as when 5 men were working. Consequently:

$$6 \times \frac{3}{5} = \frac{18}{5} = 3\frac{3}{5} \text{ Miles of track.}$$

Thus we know that if 3 men worked 40 days they would build 3-3/5 miles of track.

But another factor serves to lessen production. And that is the decrease in time. Only 15 days are expended, or 15/40 of the time that was expended before. Consequently:

$$3\frac{3}{5} + \frac{15}{40} = \frac{18}{5} \times \frac{15}{40}$$
$$= \frac{270}{200} = 1\frac{7}{20} \text{ Miles of track.}$$

The two arithmetical operations just shown can, of course, be combined into one. Thus:

$$6 \times \frac{3}{5} \times \frac{15}{40} = 1\frac{7}{20} \text{ Miles of track.}$$

Answer

26. If 5 men build 6 miles of railroad track in 40 days, how many miles of track can be built by 8 men working 90 days?

SOLUTION: Here we have a problem which is similar to the previous one, with this exception: more men are working a longer period of time and consequently the answer will yield not a reduced but an increased number of miles of track.

If we multiply a number by a fraction whose value is less than one, we are reducing the value of that number. If, however, we multiply the number by a fraction whose value is more than one, we are increasing the value of that number. In solving this problem, then, we would not multiply $6 \times \frac{40}{90} \times \frac{5}{8}$. That would produce a number less than 6 and we know that our answer should be more than 6 since we have more men working a longer time than were required to produce 6 miles of track. The proper way of expressing the facts given in the example is:

$$6 \times \frac{90}{40} \times \frac{8}{5} = \frac{4320}{200} = 21\frac{12}{20} \text{ Miles of track.}$$

Answer

Some examples indicate an increase in one factor and a decrease in another.

27. If 10 men earn \$500 in 12 days, how much will 6 men earn in 15 days?

SOLUTION: The number of men involved decreases and consequently the fraction will be less than one. The smaller number will therefore be the numerator.

$$\$500 \times \frac{6}{10}$$

The number of days worked increases and so the fraction will be more than one. The larger number will therefore be the numerator.

$$\$500 \times \frac{6}{10} \times \frac{15}{12} = \frac{45{,}000}{120} = \$375$$

Answer

Mixture Problems

28. A wine merchant has 32 gallons of wine worth $1.50 a gallon. If he wishes to reduce the price to $1.20 a gallon, how many gallons of water must he add?

(A) 10
(B) 9
(C) 8
(D) 7.

SOLUTION: First let us find the cost of the 32 gallons of undiluted wine at the old price.
32 × $1.50 = $48.00.

$48, then, is the value of the wine that is ultimately to be mixed with water. By the conditions of the problem, $48 will be the price realized from the sale of the wine at the new price of $1.20.

To find how many gallons of wine we shall have at the new price let us divide $48 by $1.20, the new price. The answer, of course, is 40. We see then that we must have 40 gallons of the $1.20 wine.

This is 8 gallons more than the undiluted wine. And that difference of 8 gallons is made up by water.

Answer (C) is correct.

29. A bakery shop sold 3 kinds of cake. The prices of these three kinds were 25¢, 30¢, and 35¢ per pound. The income from these sales was $36. If the number of pounds of each kind of cake sold was the same, how many pounds were sold?

SOLUTION: To buy all three kinds of cake would cost 25¢ + 30¢ + 35¢ = 90¢.

If we divide $36, the total income, by 90¢, the total price of the three kinds of cake, we will know how many times each kind of cake was sold in order to realize the $36. $\frac{\$36.00}{\$\ .90} = 40.$

Since there were just as many of each kind of cake sold, the total number of pounds sold = 40 × 3 = 120. **Answer**

30. The number of dimes in a cash register was equal to the number of quarters. There were five times as many nickels as quarters. All these coins together totalled $120. How many of each were there?

SOLUTION: Let us again make groups, this time of coins. A group will consist of five nickels + 1 quarter + 1 dime which equals 60¢. If we divide $120 by 60¢ we find that there are 200 such 60¢ groups contained in $120. Since there are 5 nickels in every one of the 200 groups we multiply 200 by 5 to find the number of nickels—1,000.
There are 200 dimes, 200 quarters and 1,000 nickels. **Answer**

FIREFIGHTER, F. D.

CORRECTNESS AND EFFECTIVENESS OF EXPRESSION

The questions contained in this section are designed to test your knowledge of grammar, sentence structure, correct usage and punctuation, and to point out some of the main sources of errors that have confounded candidates on past examinations.

FIVE TEST-TYPE QUIZZES FOR PRACTICE

Now, push forward! Test yourself and practice for your test with the carefully constructed quizzes that follow. Each one presents the kind of question you may expect on your test. And each question is at just the level of difficulty that may be expected. Don't try to take all the tests at one time. Rather, schedule yourself so that you take a few at each session, and spend approximately the same time on them at each session. Score yourself honestly, and date each test. You should be able to detect improvement in your performance on successive sessions.

A portion of the standard answer sheet is provided after each test for marking your answers in the way you'll be required to do on the actual exam. At the right of this answer sheet, to make the scoring job simpler (after you have derived your own answers), you'll find our correct answers.

ANSWER SHEETS AND CORRECT ANSWERS APPEAR AFTER EACH TEST

CORRECT USAGE TEST ONE

ERRORS IN AGREEMENT OF CASE, TENSE OR NUMBER

DIRECTIONS: In each of the following groups of sentences, select the one sentence that is grammatically INCORRECT. Mark the answer sheet with the letter of that incorrect sentence.

Correct key answers to all these test questions will be found at the end of the test.

1. (A) Everyone at camp must have his medical certificate on file before participating in competitive sports.
   (B) A crate of oranges were sent from Florida for all the children in Cabin Six.
   (C) John and Danny's room looks as if they were prepared for inspection.
   (D) Three miles is too far for a young child to walk.

2. (A) The game over, the spectators rushed out on the field and tore down the goalposts.
   (B) The situation was aggravated by disputes over the captaincy of the team.
   (C) Yesterday they lay their uniforms aside with the usual end-of-the-season regret.
   (D) It is sometimes thought that politics is not for the high-minded.

3. (A) Sandburg's autobiography, as well as his poems, are familiar to many readers.
   (B) A series of authentic records of the American Indian tribes is being published.
   (C) The Smokies is the home of the descendants of this brave tribe.
   (D) Five dollars is really not too much to pay for a book of this type.

4. (A) Being tired, I stretched out on a grassy knoll.
   (B) While we were rowing on the lake, a sudden squall almost capsized the boat.
   (C) Entering the room, a strange mark on the floor attracted my attention.
   (D) Mounting the curb, the empty car crossed the sidewalk and came to rest against a building.
   (E) Sitting down, they watched him demonstrate his skill.

5. (A) The text makes the process of developing and sustaining a successful home zoo appear to be a pleasant and profitable one.
   (B) The warmth and humor, the clear characterization of the Walmsey family, which includes three children, two dogs and two cats, is such fun to read that this reviewer found herself reading it all over again.
   (C) You will be glad, I am sure, to give the book to whoever among your young friends has displayed an interest in animals.
   (D) The consensus among critics of children's literature is that the book is well worth the purchase price.

6. (A) Not one in a thousand readers take the matter seriously.
   (B) Let it lie there.
   (C) You are not so tall as he.
   (D) The people began to realize how much she had done.
   (E) He was able partially to accomplish his purpose.

7. (A) In the case of members who are absent, a special letter will be sent.
   (B) The visitors were all ready to see it.
   (C) I like Burns's poem "To a Mountain Daisy."
   (D) John told William that he was sure he had seen it.
   (E) Both men are Yale alumni.

8. (A) B. Nelson & Co. has a sale of dacron shirts today.
   (B) Venetian blinds—called that although they probably did not originate in Ven-

ice—are no longer used as extensively as they were at one time.

(C) He determined to be guided by the opinion of whoever spoke first.

(D) There is often disagreement as to whom is the better Shakespearean actor, Evans or Gielgud.

9. (A) It did not take him long to develop an interest in the great American pastime—baseball.

(B) If you had made your way to the Whipsnade Zoo, you would have had an opportunity of seeing wild animals in more or less natural habitats.

(C) How I should have liked to have spent a few more days in Paris!

(D) Neither baseball pools nor any other form of gambling is allowed in or near the school.

10. (A) If the bill were introduced, it would provoke endless debate.

(B) Since George, with his two dogs, is to be with us, it might be better to rent a cabin.

(C) He, not I, is the one to decide.

(D) He is, however, one of those restless people who never seems content in his present environment.

11. (A) Everyone can have a wonderful time in New York if they will just not try to see the entire city in one week.

(B) Being a stranger in town myself, I know how you feel.

(C) New York is a city of man-made wonders, as awe-inspiring as those found in nature.

(D) He felt deep despair (as who has not?) at the evidence of man's inhumanity to man.

12. (A) A clerk should be careful as well as punctual, even though he or she are otherwise efficient.

(B) Regardless of whether it may be true, some students are not very studious.

(C) Not every writer can say that his opinion is always the best.

(D) We often think of people who assume airs as being affected.

13. (A) We were not entirely dissatisfied with the conditions.

(B) There is a chair and a table in the room.

(C) He was among the few writers who do not experience difficulty with words.

(D) Remember that the man to whom you are speaking is my friend.

Correct Answers For The Foregoing Questions

To assist you in scoring yourself we have provided Correct Answers alongside your Answer Sheet. May we therefore suggest that while you are doing the test you cover the Correct Answers with a sheet of white paper.....to avoid temptation and to arrive at an accurate estimate of your ability and progress.

Answer Sheet

	A	B	C	D	E
1					
2					
3					
4					
5					
6					
7					
8					
9					
10					
11					
12					
13					

SCORE

..................%

NO. CORRECT ÷ NO. OF QUESTIONS ON THIS TEST

Correct Answers

	A	B	C	D	E
1		▮			
2			▮		
3	▮				
4			▮		
5		▮			
6	▮				
7				▮	
8				▮	
9			▮		
10				▮	
11	▮				
12	▮				
13		▮			

CORRECT USAGE TEST TWO

In each of the following groups of sentences, select the one sentence which is grammatically INCORRECT.

ANSWER SHEET AND CORRECT ANSWERS FOLLOW LAST QUESTION.

ERRORS IN AGREEMENT OF CASE, TENSE OR NUMBER

1. (A) This is the woman whom I saw.
   (B) She could solve even this problem.
   (C) She divided the money among the three of us.
   (D) Either she or I are guilty.

2. (A) Consider that the person which is always idle can never be happy.
   (B) Because a man understands a woman does not mean they are necessarily compatible.
   (C) He said that accuracy and speed are both essential.
   (D) Can it be said that the better of the two books is less expensive?

3. (A) Neither the critics nor the author were right about the reaction of the public.
   (B) The senator depended upon whoever was willing to assist him.
   (C) I don't recall any time when Edgar has broken his word.
   (D) Everyone of the campers but John and me is going on the hike.

4. (A) Everyone entered piomptly but her.
   (B) Each of the messengers were busily occupied.
   (C) At which exit did you leave him?
   (D) The work was not done well.

5. (A) Never before have I seen anyone who has the skill John has when he repairs engines.
   (B) If anyone can be wholly just in his decisions, it is he.
   (C) Because of his friendliness, the new neighbor was immediately accepted by the community.
   (D) Imagine our embarrassment when us girls saw Miss Maltinge sitting with her beau in the front row.

6. (A) I wondered why it was that the Mayor objected to the Governor's reference to the new tax law.
   (B) I have never read "Les Miserables," but I plan to do so this summer.
   (C) After much talk and haranguing, the workers received an increase in wages.
   (D) The author and myself were the only gay ones at the macabre gathering.

7. (A) The doctor had carelessly left all the instruments on the operating table.
   (B) Was it them whom the professor regarded with such contempt?
   (C) Despite all the power he has, I should still hate to be in his shoes.
   (D) I feel bad because I gave such a poor performance in the play tonight.

8. (A) It seems to me that everyone of the waiters is well able to do his share of the work.
   (B) His parents told him to wait at the station for his uncle and to bring him home in the car.
   (C) Who did you think was going to win the tennis match?
   (D) If he would have answered the detective honestly, he would not have been arrested.

9. (A) The general regarded whomever the colonel honored with disdain.
   (B) Everyone who reads this book will think themselves knights errant on missions of heroism.

(C) The reason why the new leader was so unsuccessful was that he had fewer responsibilities.

(D) All the new mechanical devices we have today have made our daily living a great deal simpler, it is said.

10. (A) The town consists of three distinct sections, of which the western one is by far the larger.

(B) Of London and Paris, the former is the wealthier.

(C) Chicago is larger than any other city in Illinois.

(D) America is the greatest nation, and of all other nations England is the greatest.

11. (A) I can but do my best.

(B) I cannot help comparing him with his predecessor.

(C) I wish that I was in Florida now.

(D) I like this kind of grapes better than any other.

12. (A) Neither Tom nor John was present for the rehearsal.

(B) The happiness or misery of men's lives depends on their early training.

(C) Honor as well as profit are to be gained by these studies.

(D) The egg business is only incidental to the regular business of the general store.

13. (A) It was superior in every way to the book previously used.

(B) His testimony today is different from that of yesterday.

(C) If you would have studied the problem carefully you would have found the solution more quickly.

(D) The flowers smelled so sweet that the whole house was perfumed.

14. (A) When either or both habits become fixed, the student improves.

(B) Neither his words nor his action was justifiable.

(C) A calm almost always comes before a storm.

(D) The gallery with all its pictures were destroyed.

Correct Answers For The Foregoing Questions

(Please make every effort to answer the questions on your own before looking at these answers. You'll make faster progress by following this rule.)

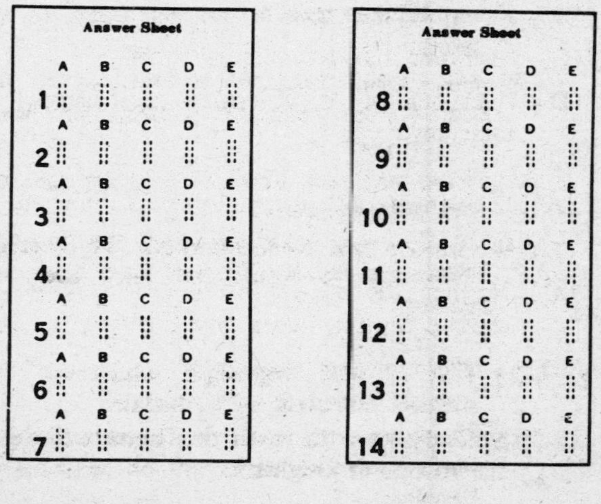

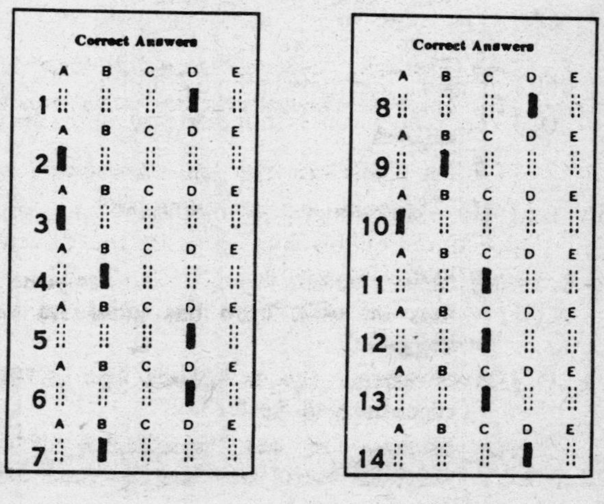

CORRECT USAGE TEST THREE

In each of the following groups of sentences, select the one sentence which is grammatically INCORRECT.

ANSWER SHEET AND CORRECT ANSWERS FOLLOW LAST QUESTION.

ERRORS IN AGREEMENT OF CASE, TENSE OR NUMBER

1. (A) Who did they say won?
   (B) The man whom I thought was my friend deceived me.
   (C) Send whoever will do the work.
   (D) The question of who should be leader arose.

2. (A) A box of choice figs was sent him for Christmas.
   (B) Neither Charles nor his brother finished his assignment.
   (C) There goes the last piece of cake and the last spoonful of ice cream.
   (D) Diamonds are more desired than any other precious stones.

3. (A) As long as you are ready, you may as well start promptly.
   (B) My younger brother insists that he is as tall as me.
   (C) We walked as long as there was any light to guide us.
   (D) Realizing I had forgotten my gloves, I returned to the theatre.

4. (A) The supervisor has given you and me difficult assignments.
   (B) This is just between you and I.
   (C) That question is not for you and me to answer.

5. (A) His knowledge of methods and procedures enable him to assist the director in many ways.
   (B) A new set of rules and regulations has been made.
   (C) Reports that the strike has been settled were circulated yesterday.

6. (A) The Credit Bureau rates you as high as him.
   (B) He is no better than you or me.
   (C) You will be notified as soon as I.

7. (A) Neither the stenographer nor the typist has returned from lunch.
   (B) Either the operators or the machines are at fault.
   (C) One or the other of those clerks are responsible for these errors.

8. (A) The Board of Directors has prepared a manual for their own use.
   (B) The company has announced its new policy of advertising.
   (C) The jury were out about thirty minutes when they returned a verdict.

9. (A) Who does he think he is?
   (B) Whom does he consider in making a decision?
   (C) Whom did they say is to be appointed?

10. (A) Who shall I say called?
    (B) The water has frozen the pipes.
    (C) Everyone has left except them.
    (D) Everyone of the salesmen must supply their own car.

11. (A) Two-thirds of the building is finished.
    (B) Where are Mr. Keene and Mr. Herbert?
    (C) Neither the salesladies nor the floorwalker want to work overtime.
    (D) The committee was agreed.

12. (A) Amends have been made for the damage to one of our cars.

(B) Neither the customer nor the clerk were aware of the fire in the store.

(C) A box of spare pencils is on the desk.

(D) There is the total number of missing pens.

13. (A) The company insist on everyone's being prompt.

(B) Each one of our salesmen takes an aptitude test.

(C) It is the location that appeals to me.

(D) Most of the men have left the building.

Correct Answers For The Foregoing Questions

(Please make every effort to answer the questions on your own before looking at these answers. You'll make faster progress by following this rule.)

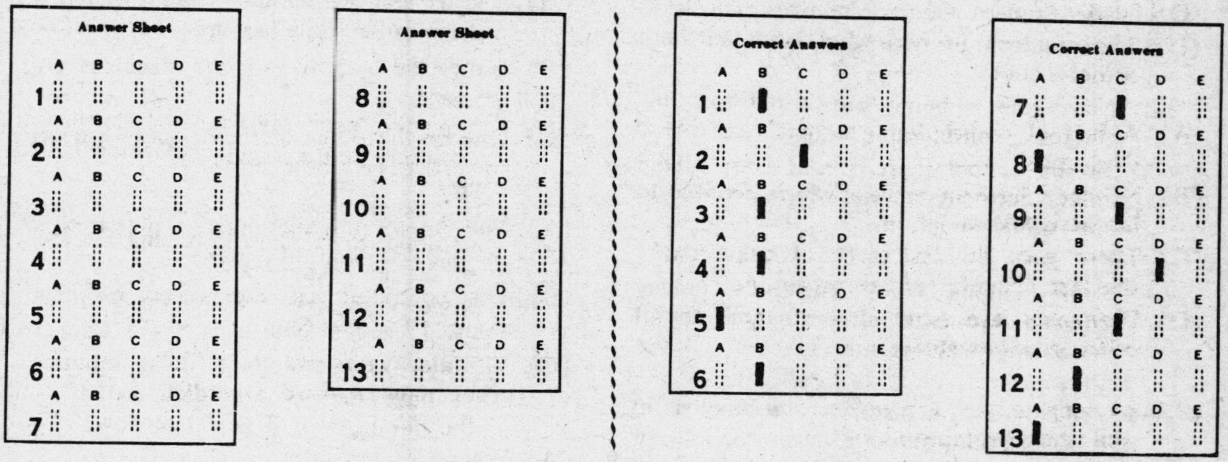

CORRECT USAGE TEST FOUR

In each of the following groups of sentences, there is one that is unacceptable because of some grammatical fault.

ANSWER SHEET AND CORRECT ANSWERS FOLLOW LAST QUESTION.

GOOD AND BAD ENGLISH

1. (A) The students in the dormitories were forbidden, unless they had special passes, from staying out after 11:30 p.m.
   (B) The Student Court rendered a decision satisfactory to both the defendant and the accuser.
   (C) Margarine is being substituted for butter to a considerable extent.
   (D) In this school there are at least fifteen minor accidents a year which are due to this traffic violation.

2. (A) Sailing along New England's craggy coastline, you will relive a bygone era of far-roving whalers and graceful clipper ships.
   (B) The march of history is reenacted in folk festivals, outdoor pageants, and fiestas—local in theme, but national in import.
   (C) Visiting the scenes of the past, our interest in American history is renewed and enlivened.
   (D) What remained was a few unrecognizable fragments.

3. (A) I knew it to be him by the style of his clothes.
   (B) No one saw him doing it.
   (C) Her going away is a loss to the community.
   (D) Mary objected to her being there.
   (E) Illness prevented him graduating in June.

4. (A) No one but her could have recognized him.
   (B) She knew the stranger to be him whom she had given up as lost.
   (C) He looked like he had been in some strange land where age advanced at a double pace.
   (D) It is impossible to include that item; the agenda have already been mimeographed.

5. (A) You have probably heard of the new innovation in the regular morning broadcast.
   (B) During the broadcast you are expected to stand, to salute, and to sing the fourth stanza of "America."
   (C) None of the rocks which form the solid crust of our planet is more than two billion years old.
   (D) "I have finished my assignment," said the pupil. "May I go home now?"

6. (A) The coming of peace effected a change in her way of life.
   (B) Spain is as weak, if not weaker than, she was in 1900.
   (C) In regard to that, I am not certain what my attitude will be.
   (D) That unfortunate family faces the problem of adjusting itself to a new way of life.
   (E) Fred Eastman states in his essay that one of the joys of reading lies in discovering courage.

7. (A) Participation in active sports produces both release from tension as well as physical well-being.
   (B) The problem of taxes is still with them.
   (C) Every boy and every girl in the auditorium was thrilled when the color guard appeared.

(D) At length our club decided to send two representatives to the meeting, you and me.

8. (A) Remains of an ancient civilization were found near Mexico City.
   (B) It is interesting to compare the interior of one of the pyramids in Mexico to the interior of one of the pyramids in Egypt.
   (C) In two days' journey you will be reminded of political upheavals comparable to the volcanic eruptions still visible and audible in parts of Mexico.
   (D) There is little danger of the laws being broken, so drastic is the penalty.

9. (A) Instead of looking disdainfully at London grime, think of it as a mantle of tradition.
   (B) Nobody but the pilot and the co-pilot was permitted to handle the mysterious package.
   (C) Not only is industry anxious to hire all available engineers, but they are being offered commissions by the armed forces.
   (D) For immediate service go direct to the store manager.

10. (A) The delegates alighted and started off in a taxi. their baggage having been taken care of.
    (B) That kind of potato is grown in Idaho.
    (C) Besides Alan Stevens, there were eight officers of the organization on the dais.
    (D) As the delegates reached the convention hall late, they blamed their tardiness on the taxi driver.

11. (A) The new system is superior from every point of view to the inefficient system in use until now.
    (B) The reason for the strike, you may recall, was because the union demanded a closed shop.
    (C) Who's to decide whether it is to be installed?
    (D) To suit Mr. Knolls, the new device will have to save time, money, and the dispositions of the employees.

Correct Answers For The Foregoing Questions

To assist you in scoring yourself we have provided Correct Answers alongside your Answer Sheet. May we therefore suggest that while you are doing the test you cover the Correct Answers with a sheet of white paper.....to avoid temptation and to arrive at an accurate estimate of your ability and progress.

Answer Sheet

	A	B	C	D	E
1					
2					
3					
4					
5					
6					
7					
8					
9					
10					
11					

Correct Answers

	A	B	C	D	E
1	▮				
2			▮		
3					▮
4			▮		
5	▮				
6		▮			
7	▮				
8		▮			
9			▮		
10				▮	
11		▮			

SCORE

%

NO. CORRECT ÷ NO. OF QUESTIONS ON THIS TEST

CORRECT USAGE TEST FIVE

GOOD AND BAD ENGLISH

DIRECTIONS: In each of the following groups of sentences, select the one sentence that is grammatically INCORRECT. Mark the answer sheet with the letter of that incorrect sentence.

Correct key answers to all these test questions will be found at the end of the test.

1. (A) The zinnia has the more vivid color, but the violet is the sweeter-smelling.
   (B) About three-fourths of the review I read was merely a summary of the story; the rest, criticism.
   (C) I shall insist that he not be accepted as a member, since he is very bad-tempered.
   (D) No sooner had he begun to speak when his auditors started to boo and hiss.

2. (A) The children's determination to find their dog almost resulted in tragedy.
   (B) They spent the first night in a house that was unlocked and with no one at home.
   (C) "What he asked me," said the boy, "was, 'Where can I find your father!'"
   (D) It was the whimpering of the younger child and the comforting words of her brother that a member of the search-party heard about ten feet off the road.

3. (A) This is between you and me.
   (B) These questions are less difficult than those.
   (C) The former is considerably different with the latter.
   (D) We are not unaware of the consequences.

4. (A) This person also wished to state his case.
   (B) He came almost immediately.
   (C) Returning to the room, the book was missing.
   (D) They are both very nearly alike in this respect.

5. (A) What kind of a job have you?
   (B) He did exactly as he was told.
   (C) The red house is differently shaped than the blue.
   (D) We hoped to see you there.

6. (A) He found neither the money or the papers.
   (B) There are no two men in the room.
   (C) Who would file the summons was the question.
   (D) It is possible he may be here now.

7. (A) The choice finally remained between them and us.
   (B) Is this the gardener of whom you spoke?
   (C) The thieves divided the money among themselves.
   (D) Do you know who this package is to be addressed to?

8. (A) I shouldn't have done it.
   (B) After a wearying day, he lay down to rest.
   (C) These problems are difficult as the others.
   (D) She saw neither of us.

9. (A) The incident occurred when he took the book off the desk.
   (B) Helen thinks as clearly as John.
   (C) I cannot believe but what he is guilty.
   (D) He indicated that the two boys were to go.

10. (A) Our child will be successful without any assistance from either you or me.
    (B) Whom do you think we ought to consider for the chairmanship?

(C) Deep in the forest, warmed by the dying embers of our camp-fire, sat the scoutmaster and I.

(D) The part of the story I enjoyed most was where the heroine decided to leave her husband.

11. (A) "Which is the way to Paris?" asked the American soldier glumly.
    (B) The physicist muttered in despair, "Will I always be surrounded by fools?"
    (C) I never feel badly if after trying hard I fail to win a prize; the effort gives me satisfaction.
    (D) Everywhere I go, I find grime and dirt in the air.

12. (A) Parents are the ones who we believe ought to insist upon their children's obeying orders.
    (B) The carpenters were lying on the ground and resting when I came in.
    (C) I do not understand why mother should object to me playing the piano at the party.
    (D) Lie in bed awhile until your aches and pains subside.

13. (A) She avoided my look of surprise, staring at the ceiling steadily.
    (B) When I told her I was sorry, I partly meant it.
    (C) My father enjoys fresh air, sunshine, and to take long walks.
    (D) The greatest pleasure I get is to see my favorite team win a ball game.

14. (A) Students who plan to become physicians are advised to study biology, chemistry, and German.
    (B) The critic asked the author which of his plays he considered the best.
    (C) I, who am your best friend, should have at least a fair chance of winning the prize.
    (D) If you go past the library tomorrow, please bring this book to the librarian who sits at the desk in the children's room.

15. (A) Writers no longer take for granted the mores of the society in which they live.
    (B) He is in this country now for five years, but he makes no attempt to speak our language.
    (C) The reason the child rebelled was that the order made no sense to him.
    (D) His written work has been done in so careless a manner that I refuse to read it.

Correct Answers For The Foregoing Questions

(Please make every effort to answer the questions on your own before looking at these answers. You'll make faster progress by following this rule.)

Answer Sheet

	A	B	C	D	E
1					
2					
3					
4					
5					
6					
7					
8					

Answer Sheet

	A	B	C	D	E
9					
10					
11					
12					
13					
14					
15					

Correct Answers

	A	B	C	D	E
1				▮	
2		▮			
3			▮		
4			▮		
5	▮				
6	▮				
7				▮	

Correct Answers

	A	B	C	D	E
8			▮		
9			▮		
10				▮	
11			▮		
12			▮		
13			▮		
14				▮	
15		▮			